A Technical Introduction to digital video

A Technical Introduction to

digital video

Charles A. Poynton

John Wiley & Sons, Inc.

New York • Chichester • Brisbane • Toronto • Singapore

Publisher: Katherine Schowalter
Senior Editor: Diane D. Cerra
Managing Editor: Mark Hayden
Design, illustration, and typography: Charles A. Poynton

This text is printed on acid-free paper.

Library of Congress Cataloging-in-Publication Data

Poynton, Charles A. 1950–

A Technical Introduction to Digital Video / Charles A. Poynton

352 p. 19× 23.5 cm

Includes index.

ISBN 0-471-12253-X (cloth: alk. paper)

1. Digital video. 2. Digital television.
I. Title.

TK6680.5.P67 1996

621.388—dc20 95-38474

Printed in the United States of America

10 9 8 7 6 5 4 3 2 1

to Alana and Brenna

Acknowledgments

My introduction to digital video was writing microcode for hardware conceived by John Lowry and engineered by Richard Kupnicki. I thank them both. John Ross of Ross Video continued my education in video.

I spent many hours at CJOH-TV in Ottawa, Canada, testing my theories at the invitation of CJOH's Vice-President of Engineering, Austin Reeve. I thank him for his confidence, good humor, and patience.

LeRoy DeMarsh, Junji Kumada, Ed Giorgianni, and Bill Cowan each spent many hours teaching me about color.

I thank Andreas Bechtolsheim, John Gage, Paul Borrill, and Linda Bohn of Sun Microsystems Computer Corporation. All four gave me quite a bit of latitude to define my job at Sun. Susan Hathaway taught me about writing and editing. Jayna Pike helped me understand how to produce a book. Nick England, Kipp Kramer, and Bill Pratt suffered through some turbulent times at Sun, and helped me through some of mine.

I thank Eileen McGinnis, Lisa Leffingwell, Mary Medoff, Foster Roizen, and John Streets for their friendship, especially during my stay in Palo Alto.

Diane Cerra is my editor; she is assisted by Tammy Boyd. It is a pleasure to work with them, and with my production editor Mark Hayden.

Much of the material in this book originated from my courses, seminars, and tutorials. I have benefited from the patience and organizational talent of two people who have invited me to present seminars: Linda Young of DuArt Film & Video, and Charles Swartz of UCLA. I am also grateful for the contributions of students who have attended my SIGGRAPH courses, particularly Mary Mooney.

I thank the colleagues who encouraged me in this project, some of whom reviewed the manuscript at various stages: Michael Bourgoin, John Galt, George Joblove, Keiichi Kubota, Ken R. Leese, Barry P. Medoff, Cam Morrison, Thor Olson, Glenn A. Reitmaier, Laurence J. Thorpe, and especially Fred Remley, Peter D. Symes, and Marceli Wein.

I owe a debt of gratitude to three colleagues whom I value as close personal friends: C. R. Caillouet, Pierre DeGuire, and Charlie Pantuso. All three contributed greatly to my understanding of video.

Thanks, Mom! Thanks, Dad! Thanks, Al and Peg!

Finally, and most of all, I thank Leigh Davidson.

Contents

Appendix

List of figures

List of tables

Preface

A year ago, among the exhibits of a trade show, I had a conversation with a design manager for a firm that designs and manufactures integrated circuits for MPEG compression. He said something that amazed me: during development of their latest MPEG decoder chip, he spent a lot of time "tweaking the coefficients of the color decoding matrices to get the picture to look right."

This statement amazed me because if the circuits are designed properly, no tweaking is necessary! There are subjective aspects in image capture and reproduction, but they do not extend to the matrix coefficients. I concluded that he did not have access to information that, first, would have given him an understanding of the processes involved in video encoding and decoding and, second, would have provided the correct coefficients.

I hope this book meets both those needs.

I have left two important topics to future volumes: compression – particularly MPEG – and high definition television (HDTV). This book describes the fundamental technology that lies at the heart of both of these areas, so serves as a prerequisite.

Formulas

It is said that every formula in a book cuts the potential readership in half. I hope readers of this book can compute that after a mere ten formulas my readership would drop to 2^{-10}! I decided to retain formulas, but they are not generally necessary if you wish to understand just the concepts. If you are intimidated by a formula, just skip it and come back later if you wish.

Luma vs Luminance

Confusion between the *luminance* used in color science and the so-called luminance used in video is rampant; one highly regarded computer graphics text incorrectly declares "luminance" in video to be a linear quantity. The confusion between these interpretations is responsible for much difficulty in the exchange of digital imagery between computer graphics, video, and prepress. When I write Y' with a prime, that is video *luma*, a nonlinear quantity. When I write Y without a prime, that is the linear CIE *luminance* of color science. When you see someone else write Y without a prime, you must explore further to see which interpretation the author intends.

It is customary in video to denote color difference signals using lowercase subscripts *b* and *r*. I consider these to be at great risk of loss in reproduction, so I set the subscripts in uppercase: $Y'C_BC_R$, $Y'P_BP_R$. This introduces no ambiguity.

CCIR vs ITU-R

The CCIR has historically been the international standards-setting body for radio and through succession, television. I was bewildered in 1993 when the organization changed its name to ITU-R. Its standards, which used to be known by titles such as "CCIR Recommendation 601" suddenly became such unpronounceable items as "Rec. ITU-R BT.601." I retain the venerable Rec. designation to avoid burdening you with this clumsy nomenclature.

Standards

Consumer electronic equipment is manufactured in huge unit volumes. Studio video equipment is very expensive, and tends to have a fairly long life of about a decade. These facts impose a heavy burden of reverse compatibility on video technology, so its present state is complicated. Also, the last few years have seen several failures of the industry to standardize in a timely manner, so a certain amount of chaos has been introduced. If you are bewildered by the apparent chaos, you should know the important strategic directions:

- Rec. 709 chromaticity and transfer function,
- Rec. 601 luma coefficients and $Y'C_BC_R$ coding,
- analog interface with zero setup, 300 mV sync, and 700 mV video.

Layout and typography

I have been frustrated for many years with technical books having awkwardly placed figures. I believe there are two causes for this. First, technical authors tend to have little feeling for visual structure, and little background or experience in drawing. Second, the writing, illustration, and layout of a technical book are usually executed by different people working independently. Even if the obvious problem of having a figure located one or two pages away from its reference is avoided, the lack of coordination of text and illustration is felt by the reader as a jerkiness. I have attempted to avoid this situation by executing my own typesetting.

Robert Bringhurst, *The Elements of Typographic Style*. Vancouver, BC: Hartley & Marks, 1992.

Jan Tschichold, *The Form of the Book*. London: Lund Humphries, 1991 (originally published in German in 1975).

Edward R. Tufte, *Envisaging Information*. Cheshire, CT: Graphic Press, 1990.

In designing and typesetting this book, I was inspired by the work of three people: Robert Bringhurst, Jan Tschichold, and Edward R. Tufte.

Further reading

Rudolf Mäusl, *Refresher topics: television technology.* Munich: Rohde & Schwarz, 1990, PD 756.8990.21.

I have been inspired for many years by the concise, accurate, and authoritative booklet by Prof. Rudolf Mäusl, issued by Rohde & Schwarz.

K. Blair Benson, *Television Engineering Handbook, Featuring HDTV Systems, Revised Edition,* revised by Jerry C. Whitaker. New York, McGraw-Hill, 1992. This supersedes the Second Edition.

Television Engineering Handbook is an exhaustive reference text once you are familiar with the topic, although it suffers from the lack of cohesion typical of a 1500-page compilation.

C. P. Sandbank, Editor, *Digital Television.* Chichester, England: John Wiley & Sons, 1990.

C. P. Sandbank has edited an excellent, in-depth work on the technical aspects of digital television, concentrating on signal processing, recording, and transmission. I recommend it.

Basic principles 1

This chapter is a summary of the fundamental concepts of digital video.

If you are unfamiliar with video, this chapter will introduce the major issues, to acquaint you with the framework and nomenclature that you will need to address the rest of the book. If you are already knowledgeable about video, this chapter will provide a quick refresher, and will direct you to specific topics about which you'd like to learn more.

Imaging

The three-dimensional world is imaged by the lens of the human eye onto the retina, which is populated with photoreceptor cells that respond to light having wavelengths in the range of about 400 nm to 700 nm. In an imaging system, we build a camera having a lens and a photosensitive device, to mimic how the world is perceived by vision.

Although the shape of the retina is roughly a section of a sphere, it is topologically two-dimensional. In a camera, for practical reasons, we employ a flat *image plane,* sketched in Figure 1.1 overleaf, instead of a spherical image surface. Image system theory concerns analyzing the continuous distribution of power that is incident on the image plane.

A photographic camera has, in the image plane, film that is subject to chemical change when irradiated by

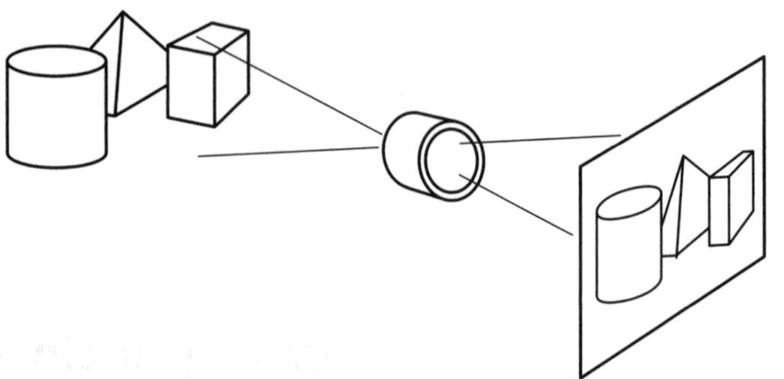

Figure 1.1 **Scene, lens, image plane.**

light. The active ingredient of photographic film is contained in a thin layer of particles having carefully controlled size and shape, in a pattern with no coherent structure. If the particles are sufficiently dense, an image can be reproduced that has sufficient information for a human observer to get a strong sense of the original scene. The finer the particles and the more densely they are arranged in the film medium, the higher will be the capability of the film to record spatial detail.

Digitization

Signals captured from the physical world are translated into digital form by *digitization*, which involves two processes. A signal is digitized when it is subjected to both *sampling* and *quantization*, in either order. When an audio signal is sampled, the single dimension of time is carved into discrete intervals. When an image is sampled, two-dimensional space is partitioned into small, discrete regions. Quantization assigns an integer to the amplitude of the signal in each interval or region.

1-D sampling

A signal that is a continuous one-dimensional function of time, such as an audio signal, is sampled through forming a series of discrete values, each of which represents the signal at an instant of time. *Uniform sampling*, where the time intervals are of equal duration, is ubiquitous.

2-D sampling

A continuous two-dimensional function of space is sampled by assigning, to each element of a sampling grid, a value that is a function of the distribution of intensity over a small region of space. In digital video and in conventional image processing, the samples lie on a regular, rectangular grid.

Samples need not be digital: A CCD camera is inherently sampled, but it is not inherently quantized. Analog video is not sampled horizontally, but is sampled vertically by scanning, and sampled temporally at the frame rate.

Pixel array

A digital image is represented by a matrix of values, where each value is a function of the information surrounding the corresponding point in the image. A single element in an image matrix is a *picture element*, or *pixel*. In a color system, a pixel includes information for all color components. Several common formats are sketched in Figure 1.2 below.

In computing it is conventional to use a sampling grid having equal horizontal and vertical sample pitch – *square pixels*. The term *square* refers to the sample pitch; it should not be taken to imply that image information associated with the pixel is distributed uniformly throughout a square region. Many video systems use sampling grids where the horizontal and vertical sample pitch are not equal.

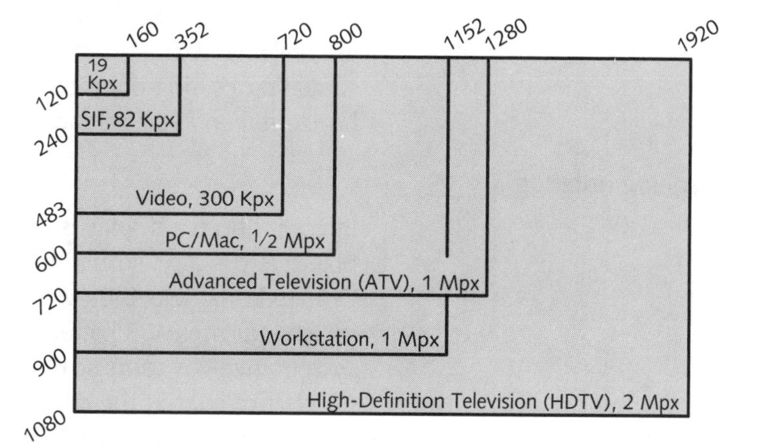

Figure 1.2 **Pixel array.**

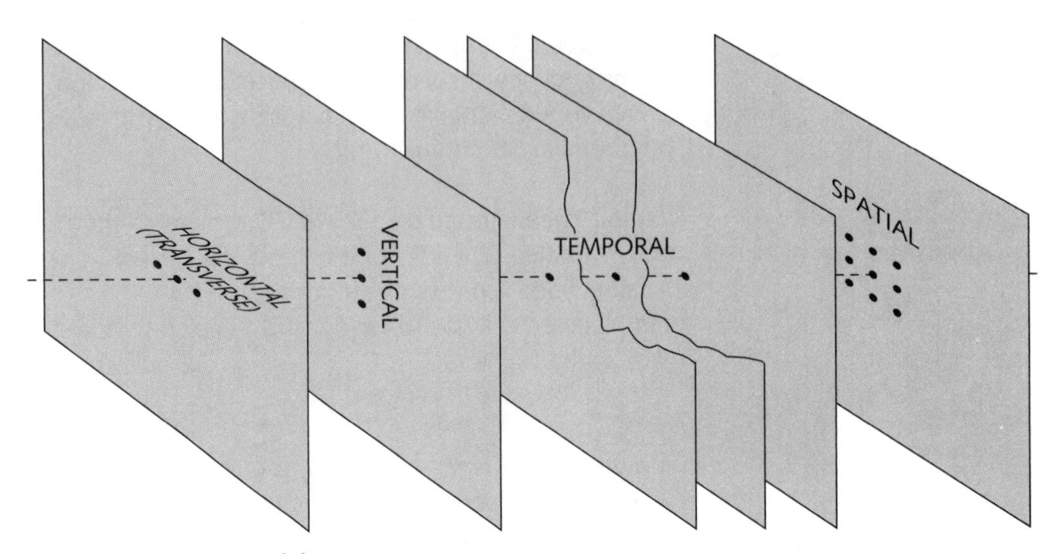

Figure 1.3 **Spatiotemporal domains.**

Some framebuffers provide a fourth byte, which may be unused, or used to convey overlay or transparency data.

In computing it is usual to represent a grayscale or pseudocolor pixel as a single 8-bit byte. It is common to represent a truecolor pixel as three 8-bit red, green, and blue (*R'G'B'*) components totaling three bytes – 24 bits – per pixel.

Spatiotemporal domains

A digital video image is sampled in the horizontal, vertical, and temporal axes, as indicated in Figure 1.3 above. One-dimensional sampling theory applies along each of these axes. At the right is a portion of the two-dimensional *spatial* domain of a single image. Some spatial processing operations cannot be separated into horizontal and vertical facets.

Scanning notation

In computing, a display is described by the count of pixels across the width and height of the image. Conventional television would be denoted 644×483, which indicates 483 picture lines. But any display system involves some scanning overhead, so the total number of lines in the *raster* of conventional video is necessarily greater than 483.

Video scanning systems have traditionally been denoted by their total number of lines including sync and blanking overhead, the frame rate in hertz, and an indication of *interlace* (2:1) or *progressive* (1:1) scan, to be introduced on page 11.

525/59.94/2:1 scanning

is used in North America and Japan, with an analog bandwidth for studio video of about 5.5 MHz.

625/50/2:1 scanning

is used in Europe and Asia, with an analog bandwidth for studio video of about 6.5 MHz. For both 525/59.94 and 625/50 component digital video according to ITU-R Rec. BT.601-4 ("Rec. 601"), the basic sampling rate is exactly 13.5 MHz. *Bandwidth* and *sampling rate* will be explained in later sections.

1125/60/2:1 scanning

is in use for *high-definition television* (HDTV), with an analog bandwidth of about 30 MHz. The basic sampling rate for 1125/60 is 74.25 MHz. A variant 1125/59.94/2:1 is in use. This scanning system was originally standardized with a 1920×1035 image having pixels about 4 percent taller than square.

1920×1080

The square-pixel version of 1125/60 is now commonly referred to as 1920×1080.

1280×720

A progressive-scan one megapixel image format is proposed for advanced television in the United States.

Viewing distance and angle

A viewer tends to position himself or herself relative to a scene so that the smallest detail of interest in the scene subtends an angle of about one minute of arc ($^1/_{60}$°), approximately the limit of angular discrimination for normal vision. For the 483 picture lines of conventional television, the corresponding viewing distance is about seven times picture height (PH); the horizontal viewing angle is about 11°. For the 1080 picture lines of HDTV, the optimum viewing distance is 3.3 screen heights, and the horizontal viewing angle is almost tripled to 28°. The situation is sketched in Figure 1.4 overleaf.

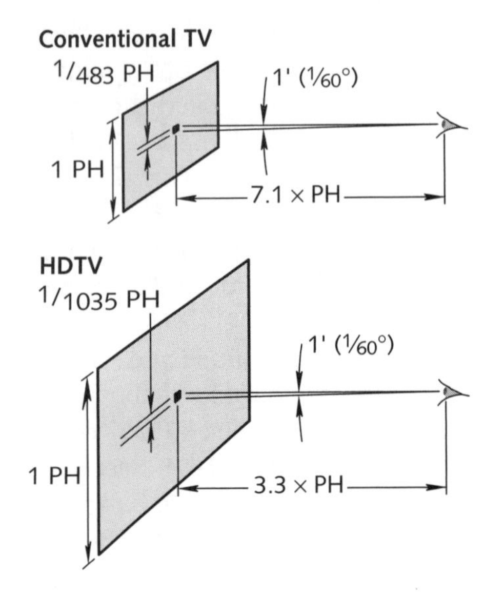

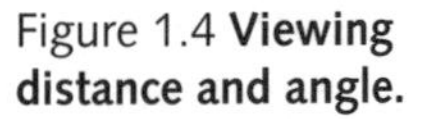

Figure 1.4 **Viewing distance and angle.**

$$distance = \frac{3400}{lines} \times PH$$

To achieve a viewing situation where a pixel subtends $1/60°$, viewing distance expressed in units of picture height should be about 3400 divided by the number of picture lines. A computer user tends to position himself or herself closer than this – about 50 to 60 percent of this distance – but at this closer distance individual pixels are discernible. Consumer projection television is viewed closer than 7×PH, but at this distance scan lines become objectionable.

Aspect ratio

Variants of conventional 525/59.94 systems having 16:9 aspect ratio have recently been standardized, but none are widely deployed as I write this.

Aspect ratio is the ratio of image width to height. Conventional television has an aspect ratio of 4:3. High-definition television uses a wider ratio of 16:9. Cinema commonly uses 1.85:1 or 2.35:1. In a system having square pixels, the number of horizontal samples per picture width is the number of scanning lines in the picture height times the aspect ratio of the image.

Frame rate, refresh rate

A succession of flashed still pictures, captured and displayed at a sufficiently high rate, can create the illusion of motion. The quality of the motion portrayal depends on many factors.

Most displays for moving images involve a period of time when the reproduced image is absent from the display, that is, a fraction of the frame time during which the display is black. In order to avoid objectionable flicker, it is necessary to flash the image at a rate higher than the rate necessary to portray motion. Refresh rate is highly dependent on the ambient illumination in the viewing environment: The brighter the environment, the higher the flash rate must be in order to avoid flicker. To some extent the brightness of the image itself influences the flicker threshold, so the brighter the image, the higher the refresh rate must be. Since peripheral vision has higher temporal sensitivity than central (foveal) vision, the flicker threshold of vision is also a function of the viewing angle of the image.

Refresh rate is generally engineered into a system. Once chosen, it cannot easily be changed. Different applications have adopted different refresh rates, depending on the image quality requirements and viewing conditions of the application.

In the darkness of a cinema, a flash rate of 48 Hz is adequate. In the early days of motion pictures, a frame rate of 48 Hz was thought to involve excessive expenditure for film stock, and 24 frames per second were found to be sufficient to portray motion. So, a conventional film projector flashes each frame twice. Higher realism can be obtained with specialized cameras and projectors that operate at higher frame rates, up to 60 frames per second or more.

In a dim viewing environment typical of television viewing, such as a living room, a flash rate of 60 Hz is sufficient. Originally, television refresh rates were chosen to match the local AC power line frequency.

In a bright environment such as an office, a refresh rate above 70 Hz might be required.

Motion portrayal

It is conventional in video for each element of an image sensor device to integrate light from the scene for the entire frame time. This captures as much of the light from the scene as possible, in order to maximize sensitivity and/or signal-to-noise ratio. In an interlaced camera, the *exposure time* is usually effectively the duration of the field, not the duration of the frame. This is necessary in order to achieve good motion portrayal.

If the image has elements that move an appreciable distance during the exposure time, then the sampled image information will exhibit *smear*. Smear can be minimized by using an exposure time that is a fraction of the frame time; however, the method involves discarding light from the scene and a sensitivity penalty is incurred.

When the effect of image information incident during a single frame time persists into succeeding frames, the sensor exhibits *lag*. Lag is a practical problem for tube-type cameras, but generally not a problem for CCD cameras.

Flicker is absent in any image display device that produces steady, unflashing light for the duration of the frame time. You might think that a nonflashing display would be more suitable than a device that flashes, and many contemporary devices do not flash. However, if the viewer's gaze is tracking an element that moves across the display, a display with an *on-time* approaching the frame time will exhibit smearing of elements that move. This problem becomes more severe as eye tracking rates increase; for example, with the wide viewing angle of high-definition television.

Raster scanning

In cameras and displays, some time is required to advance the scanning operation – to *retrace* – from one line to the next and from one frame to the next. These intervals are called *blanking intervals*, because in

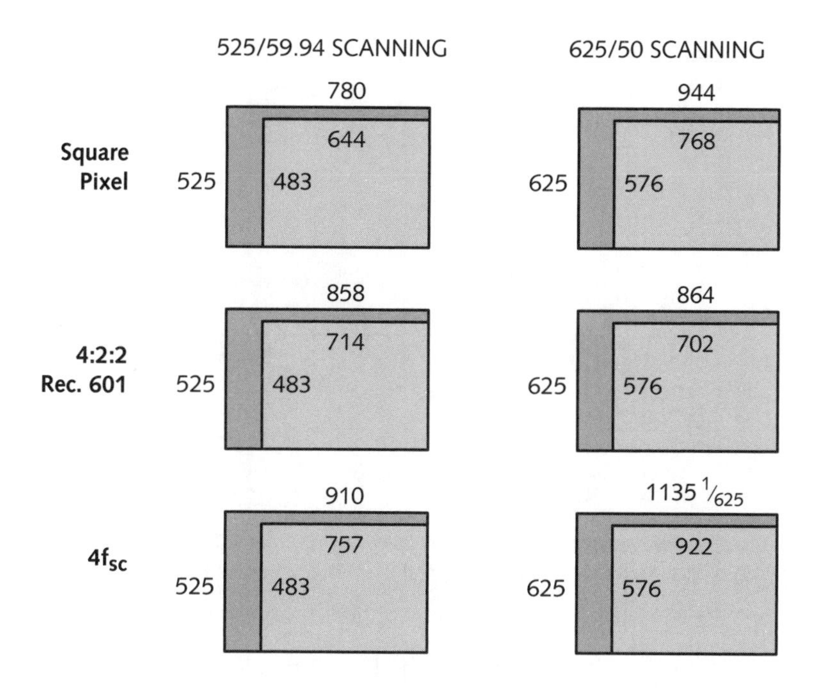

Figure 1.5 **Digital video rasters.** The left column shows 525/59.94 scanning, the right column shows 625/50. The top row shows sampling with square pixels. The middle row shows sampling at the Rec. 601 standard sampling frequency of 13.5 MHz. The bottom row shows sampling at four times the color subcarrier. Blanking intervals are shown with dark shading.

525/59.94 is colloquially referred to as *NTSC*, and 625/50 as *PAL*, but the terms NTSC and PAL properly apply to color encoding standards and not to scanning standards.

a conventional CRT display the electron beam must be extinguished (*blanked*) during these time intervals. The *horizontal blanking* time lies between scan lines, and *vertical blanking* lies between frames (or fields). Figure 1.5 above shows the raster structure of 525/59.94 and 625/50 digital video systems, including these blanking intervals. In analog video, sync information is conveyed during the blanking intervals.

The horizontal and vertical blanking intervals required for a CRT display are quite large fractions of the line time and frame time: in 525/59.94, 625/50, and 1920×1035 systems, vertical blanking occupies 8 percent of each frame period. Although in principle a digital video interface could omit the blanking intervals and use a clock having a lower frequency than the sampling clock, this would be impractical. Digital video standards use interface clock frequencies chosen to

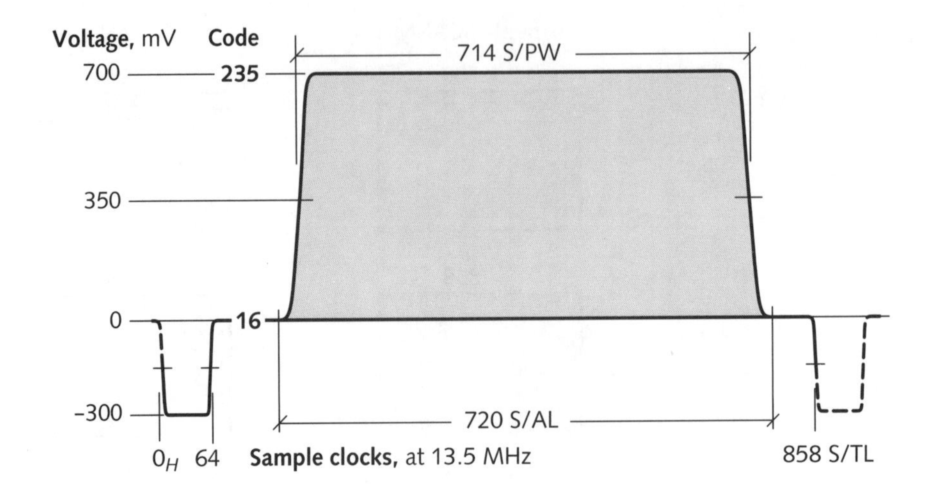

Figure 1.6 **Scan line waveform** for 525/59.94 component video, showing luma. The 720 *active* samples contain picture information. Horizontal blanking occupies the remaining sample intervals.

match the large blanking intervals of typical display equipment. Good use is made in digital systems of what would otherwise be excess data capacity: A digital video interface may convey audio signals during blanking; a digital video tape recorder might record error correction information in these intervals.

In a digital video system it is standard to convey samples of the image matrix in the same order that the image information would be conveyed in an analog video system: first the top line (left to right), then the next lower line, and so on.

In analog video, information in the image plane is scanned uniformly left to right during a fixed, short interval of time – the *active line time* – and conveyed as an analog electrical signal. There is a uniform mapping from horizontal position in the image to time instant in the electrical signal. Successive lines are scanned uniformly from the top of the image to the bottom, so there is also a uniform mapping from vertical position in the image to time instant in the electrical signal. The fixed pattern of parallel scanning lines disposed across the image is the *raster*. The word is derived from the Greek *rake*, from the resemblance of a raster to the pattern left on a newly raked field.

Figure 1.6 above shows the waveform of a single scan line, showing voltage from 0 V to 700 mV in a component analog system (with sync at -300 mV), and codeword value from code 16 to code 235 in an 8-bit component digital system.

Interlace

At the outset of television, the requirement to minimize information rate for transmission – and later, recording – led to *interlaced* scanning. Each frame is scanned in two successive vertical passes, first the *odd field*, then the *even field*, whose scan lines interlace. Total information rate is reduced because the flicker susceptibility of vision is due to a wide-area effect. As long as the complete height of the picture is scanned rapidly enough to overcome wide-area flicker, small-scale picture information – such as that in the alternate lines – can be transmitted at a lower rate. Interlaced scanning is illustrated in Figure 1.7 below.

Figure 1.7 exaggerates the slant of a fraction of a degree that results when a conventional CRT – either a camera tube or a display tube – is scanned with analog circuits. The slant is a real effect in analog cameras and displays, although it is disregarded in the design of equipment.

If the information in an image changes vertically at a scale comparable to the scanning line pitch – if a fine pattern of black-and-white horizontal line pairs is scanned, for example – then interlace can cause the content of the odd and the even fields to differ markedly. This causes *twitter*, a small-scale phenomenon that is perceived as extremely rapid back-and-forth motion. Twitter can be produced not only from degenerate images such as fine horizontal black-and-white lines, but also from high-amplitude brightness detail in an ordinary image. In computer generated imagery (CGI), twitter can be reduced by vertical filtering.

If image information differs greatly from one field to the next, then instead of twitter, large-scale flicker will

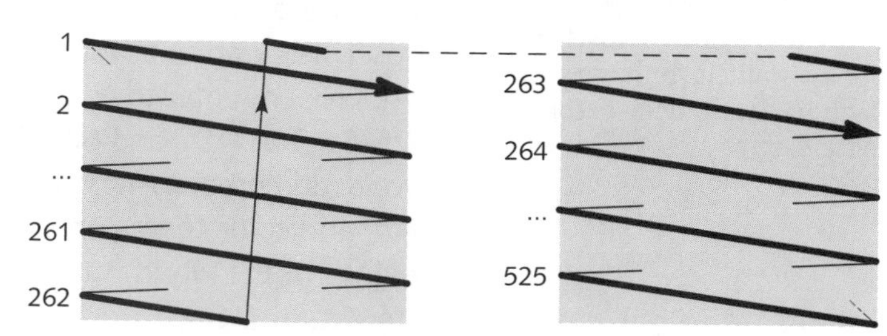

Figure 1.7 **Interlaced scanning** forms a complete picture – the *frame* – from two *fields*, each comprising half the scanning lines. The second field is delayed half the frame time from the first.

result. A video camera is designed to avoid introduction of so much vertical detail that flicker could be produced. In synthetic image generation, vertical detail may have to be explicitly filtered in order to avoid flicker.

Scanning standards

Conventional broadcast television scans a picture whose aspect ratio is 4:3, in left-to-right, top-to-bottom order using interlaced scanning.

A scanning system is denoted by its total line count and its field rate in hertz, separated by a solidus (slash). Two scanning standards are established for conventional television: 525/59.94, used primarily in North America and Japan; and 625/50, used elsewhere. It is obvious from the scanning nomenclature that the line counts and frame rates are different. There are other important differences:

525/59.94 video in Japan uses 10:4 picture to sync ratio and zero setup.

System	*525/59.94*	*625/50*
Picture:Sync ratio	10:4	7:3
Setup, percent	7.5	0
Count of equalization, broad pulses	6	5
Line number 1 defined at	First equalization pulse	First broad pulse

The two systems have gratuitous differences in other parameters unrelated to scanning.

Monochrome systems having 405/50/2:1 and 819/50/2:1 scanning were once used in Britain and France, respectively, but transmitters for these standards have now been decommissioned.

Systems with 525/59.94 scanning usually employ NTSC color coding, and systems with 625/50 scanning usually use PAL, so 525/59.94 and 625/50 systems are loosely referred to as *NTSC* and *PAL*. But NTSC and PAL properly refer to color encoding. Although 525/59.94/NTSC and 625/50/PAL systems dominate worldwide broadcasting, other combinations of scanning and color coding are in use in large and important regions of the world, such as France, Russia, and South America.

The frame rate of 525/59.94 video is exactly $^{60}/_{1.001}$ Hz. In 625/50 the frame rate is exactly 50 Hz. Computer graphics systems have various frame rates with few standards and poor tolerances.

An 1125/60/2:1 high-definition television production system has been adopted as SMPTE Standard 240M and has been proposed to the ITU-R. At the time of writing, the system is in use for broadcasting in Japan but no international broadcasting standards have been agreed upon.

All of these scanning systems are interlaced 2:1, and interlace is implicit in the scanning nomenclature. Noninterlaced scanning is common in desktop computers and universal in computer workstations. Emerging high-definition television standards have interlaced and noninterlaced variants.

John Watkinson, *The Engineer's Guide to Standards Conversion*. Petersfield, Hampshire, England: Snell & Wilcox, 1994.

Standards conversion refers to conversion among scanning standards. Standards conversion, done well, is difficult and expensive. Standards conversion between scanning systems having different frame rates, even done poorly, requires a fieldstore or framestore. The complexity of standards conversion between 525/59.94 scanning and 625/50 scanning is the reason that it is difficult for consumers – and broadcasters – to convert European material for use in North America or Japan, or vice versa.

Transcoding refers to changing the color encoding of a signal, without altering its scanning system.

Sync structure

At a video interface, synchronization (*sync*) is achieved by associating, with every scan line, a line sync datum denoted 0_H (pronounced *zero-H*). In component digital video, sync is conveyed using digital codes 0 and 255 outside the range of picture information.In analog video, sync is conveyed by voltage levels "blacker than black." 0_H is defined by the 50-percent point of the leading (falling) edge of sync.

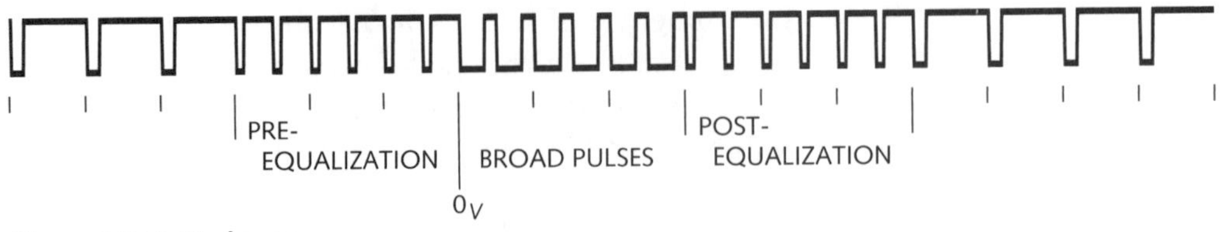

Figure 1.8 **Vertical sync waveform of 525/59.94.**

In both 525/59.94 and 625/50 video the *normal* sync pulse has a duration of 4.7 µs. Vertical sync is identified by *broad pulses*, which are *serrated* in order for a receiver to maintain horizontal sync even during the vertical interval. The start of the first broad pulse identifies the field sync datum, denoted 0_V. Narrow *equalization* pulses, half the sync pulse duration at twice the line rate, are present during intervals immediately before and immediately following the broad pulses.

These *equalization pulses* have no relationship with the process of *equalization* that is used to compensate poor frequency response of coaxial cable, or poor frequency or phase response of a filter.

When analog sync separators comprised just a few resistors and capacitors, to achieve stable interlacing required halving the duration of the line syncs and introducing additional pulses halfway between them. Originally the *equalization* pulses were the ones interposed between the line syncs, but the term now refers to all of the narrow pulses. The absence of sync level between the end of a broad pulse and the start of the following sync was called *serration*. If you think of field sync as a single pulse asserted for several lines, serration is the negation of this pulse at twice the line rate.

In digital technology it is more intuitive to consider the pulses that are present than the ones that are absent, so the term *serration* is no longer popular.

An equalization pulse has half the duration of a normal sync. The duration of a vertical (*broad*) pulse is half the line time, less a full sync width. A 525/59.94 system has three lines of *preequalization* pulses, three lines of vertical sync, and three lines of *postequalization* pulses. A 625/50 system has two and one-half lines (five pulses) of each of preequalization, broad, and post-equalization pulses. Figure 1.8 above sketches the vertical sync component of 525/59.94 analog video.

Monochrome 525-line broadcasting originated with a line rate of exactly 15.750 kHz. When color was introduced to NTSC in 1953, the monochrome horizontal frequency was multiplied by exactly $^{1000}/_{1001}$ to obtain the NTSC color line rate of approximately 15.734 kHz. Details are in *Field, frame, line, and sample rates*, on page 199. All 525-line broadcast signals – even monochrome signals – now employ this rate. The line rate of 625/50 systems has always been exactly 15.625 kHz, corresponding to a line time of exactly 64 μs.

Data rate

b = bit
B = Byte

k	10^3	1000
K	2^{10}	1024
SI, datacom:		
M	10^6	1 000 000
disk:		
M	$10^3 \cdot 2^{10}$	1 024 000
RAM:		
M	2^{20}	1 048 576

The formal, international designation of the metric system is *Système International d'Unités*, SI.

Data rate of a digital system is measured in bits per second (b/s) or bytes per second (B/s), where a byte is eight bits. The SI prefix *k* denotes 10^3 (1000); it is often used in data communications. The *K* prefix used in computing denotes 2^{10} (1024). The SI prefix *M* denotes 10^6 (1 000 000). Disk storage is generally allocated in units integrally related to 1024 bytes; the prefix *M* applied to disk storage denotes 1 024 000. RAM memory generally has capacity based on powers of two; the prefix *M* applied to RAM denotes 2^{20} or 1024 K (1 048 576).

Data rate of digital video

Line rate is an important parameter of a video system: Line rate is simply the frame rate multiplied by the number of lines per total frame.

The aggregate *data rate* is the number of bits per pixel, times the number of pixels per line, times the number of lines per frame, times the frame rate.

In both analog and digital video it is necessary to convey not only the raw image information, but also information about which time instants (or which samples) are associated with the start of frame, or the start of line. This information is conveyed by signal synchronization or *sync* elements. In analog video and composite digital video, sync is combined with video by being coded at a level *blacker than black*.

All computer graphics systems and almost all digital video systems have the same integer number of sample clock periods in every raster line. In these cases, sampling frequency is simply the line rate times the number of samples per total line (S/TL).

In 625/50 PAL there is not an exact integer number of samples per line: Samples in successive lines are offset to the left a small fraction, $1/625$ of the horizontal sample pitch. The sampling structure is not precisely *orthogonal*, although digital acquisition, processing, and display equipment treat it so.

The data capacity required for the *active* pixels of a frame is computed by simply multiplying the number of bits per pixel by the number of active pixels per line, then by the number of active lines per frame. To compute the data rate for the active pixels, simply multiply by the frame rate.

Standards are not well established in display systems used in desktop computers, workstations, and industrial equipment. The absence of published data makes it difficult to determine raster scanning parameters.

Linearity

A video system should ideally satisfy the *principle of superposition;* in other words, it should exhibit *linearity.* A function f is linear *if and only if* (iff):

Eq 1.1

$$f(a+b) \equiv f(a) + f(b)$$

The function *f* can encompass an entire system: A system is linear iff the sum of the individual responses of the system to any two signals is identical to its response to the sum of the two. Linearity can pertain to steady-state response, or to the system's temporal response to a changing signal.

Linearity is a very important property in mathematics, in signal processing, and in video. But linearity in one domain cannot be carried across to another domain if a nonlinear function separates the two. An image signal usually originates in a sensor that has linear response to physical intensity. And video signals are usually processed through analog circuits that have linear response to voltage or digital systems that are linear with respect to the arithmetic performed on the codewords. But a video camera applies a nonlinear transfer function – *gamma correction* – to the image signal. So the image signal is in a linear optical domain, and the video signal is in a linear electrical domain, but the two domains are not the same.

Perceptual uniformity

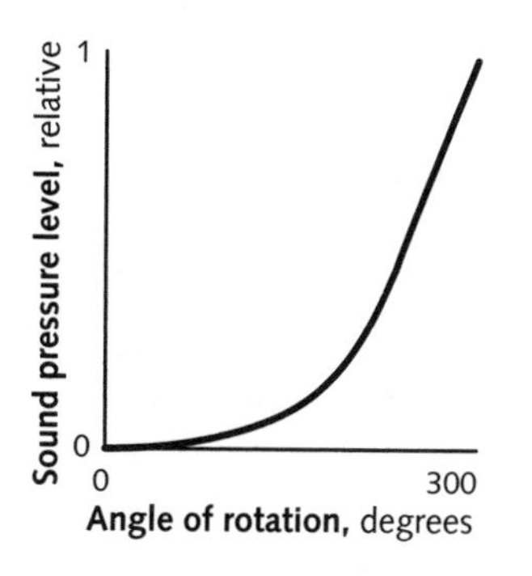

Figure 1.9 **Audio taper.**

A system is *perceptually uniform* if a small perturbation to a component value is approximately equally perceptible across the range of that value. The volume control on your radio is designed to be perceptually uniform: Rotating the knob 10 degrees produces approximately the same perceptual increment in volume anywhere across the range of the control. If the control were physically linear, the logarithmic nature of loudness perception would place all of the perceptual "action" of the control at the bottom of its range. Figure 1.9, in the margin, shows the transfer function of a potentiometer with standard *audio taper.*

The CIE L^* system, to be described on page 88, assigns a perceptually uniform scale to lightness. Video signals are coded according to perceptual principles, as will be explained in Chapter 6, *Gamma*, on page 91.

Noise, signal, sensitivity

Any analog electronic system is inevitably subject to noise that is unrelated to the signal to be processed by the system. As signal amplitude decreases, the noise makes a larger and larger relative contribution. In analog electronics, noise is inevitably introduced from thermal sources, and perhaps also from nonthermal sources of interference.

A distortion product that can be attributed to a particular processing step is known as an *artifact*, particularly if it has a distinctive visual effect on the picture.

In addition to random noise, processing of a signal may introduce distortion that is correlated to the signal itself. For the purposes of objective measurements of the performance of a system, distortion is treated as noise. Depending on its nature, distortion may be more or less perceptible than random noise.

Signal-to-Noise Ratio (SNR) is the ratio of a specified signal, often the reference amplitude or largest amplitude signal that can be carried by a system, to the amplitude of undesired components including noise and distortion. SNR is expressed in units of *decibels* (dB), a logarithmic measure.

Sensitivity refers to the minimum signal power that achieves acceptable (or specified) SNR performance.

Quantization

To make a 50-foot-long fence with fence posts every 10 feet you need six posts, not five! Take care to distinguish *levels* (here, six) from *steps* (here, five).

A signal whose amplitude takes a range of continuous values is *quantized* by assigning to each of a finite set of intervals of amplitude a discrete, numbered level. In *uniform quantization* the *steps* between levels have equal amplitude. The degree of visual impairment caused by noise in a video signal is a function of the properties of vision. In video, it is ubiquitous to digitize a signal that is a nonlinear function, usually a 0.45-power function, of physical (linear-light) intensity. The function chosen minimizes the visibility of noise.

Theoretical SNR for an n-bit quantizer:

$$20 \log_{10}\left(2^{n}\sqrt{12}\right)$$

The effect of quantizing to a finite number of discrete amplitude levels is equivalent to adding *quantization noise* to the ideal levels of a quantized signal. Quantization has the effect of introducing noise, and thereby diminishes the SNR of a digital system. Eight-bit quantization has a theoretical SNR limit of about 56 dB (peak signal to rms noise).

If an input signal has very little noise, then situations can arise when the quantized value is quite predictable at some points, but when the signal is near the edge of a quantizer step, uncertainty in the quantizer is reflected as noise. This situation can cause the reproduced image to exhibit *noise modulation*. It is beneficial to introduce roughly a quantizer step's worth of

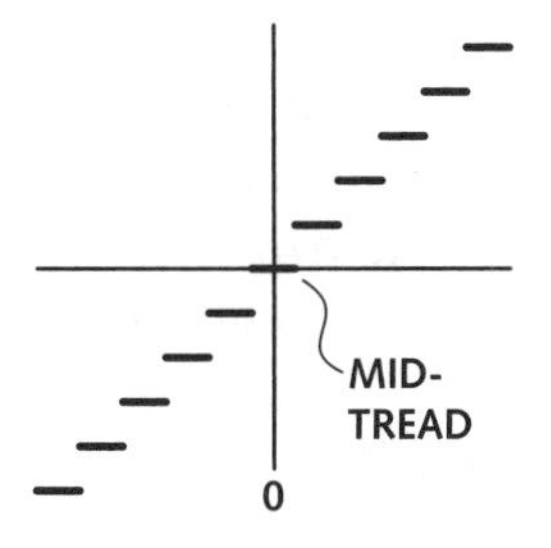

Figure 1.10
Mid-tread quantizer.

noise (peak to peak) prior to quantization, to avoid this effect. This introduces a very small amount of noise in the picture, but guarantees avoidance of "patterning" of the quantization.

Quantization can be applied to a unipolar signal such as luma. For a bipolar signal such as a color difference it is standard to use a *mid-tread* quantizer, such as the one sketched in Figure 1.10 in the margin, so that no systematic error affects the zero value.

Frequency response, bandwidth

Figure 1.11 below shows, a test signal starting at zero frequency and sweeping up to some high frequency. The response of a typical electronic system is shown in the middle graph; the response diminishes at high frequency. The envelope of that waveform – the system's *frequency response* – is shown at the bottom.

Figure 1.11 **Frequency response** of any electronic or optical system falls as frequency increases. Bandwidth is usually measured at the half-voltage point. Television displays are usually specified at *limiting resolution*, 10 percent of peak response.

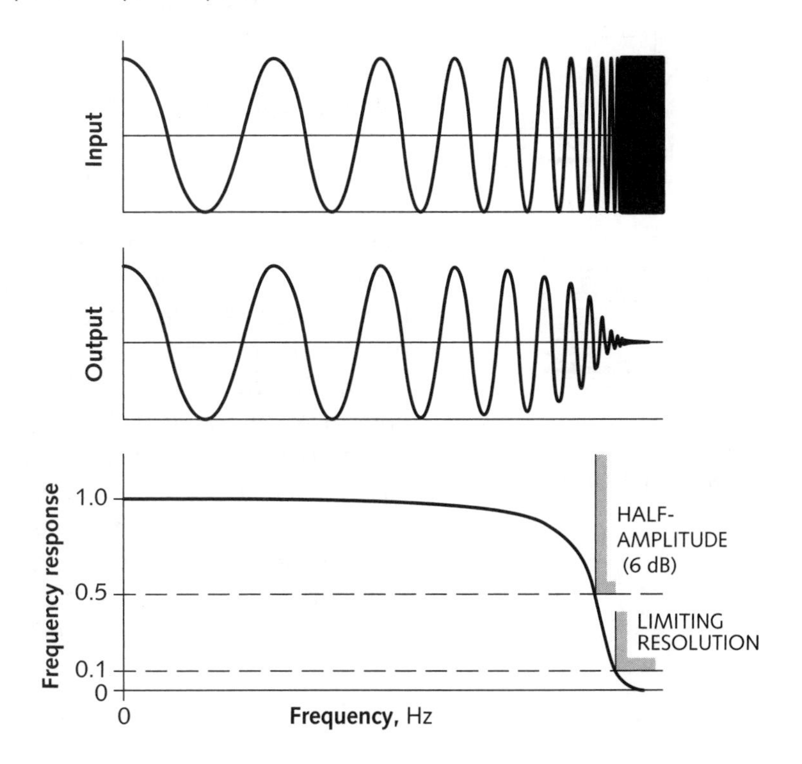

Loosely speaking, *bandwidth* is the rate at which information in a signal can change from one state to another. The response of an electronic system deteriorates above a certain information rate. Bandwidth in an analog system is generally measured at the frequency where the system's response has fallen to half of its value at a zero frequency (sometimes called *DC*).

The rate at which an analog video signal can change from one state to another say from white to black, is limited by the bandwidth of the video system. This places an upper bound on *horizontal resolution*. Consumer video generally refers to horizontal resolution, measured as the number of black and white elements (*TV lines*) that can be discerned over a horizontal distance equal to the picture height.

Bandwidth and data rate

Data rate does not apply directly to an analog system, and the term *bandwidth* does not properly apply to a digital system. When a digital system conveys a sampled representation of a continuous signal, as in digital video or digital audio, the bandwidth represented by the digitized signal is necessarily less than half – typically about 0.45 – of the sampling rate.

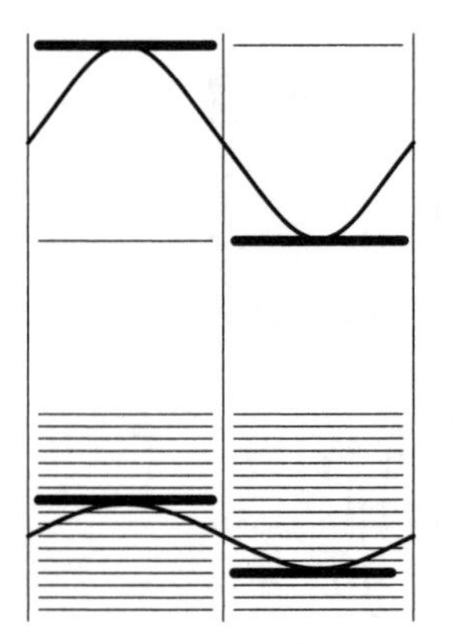

Figure 1.12
Bandwidth and data rate.

When arbitrary digital information is conveyed through an analog channel, as by a modem, the data rate that can be achieved depends on bandwidth, noise, and other properties of the channel. Figure 1.12, in the margin, shows a simple scheme that transmits two bits per second per hertz of bandwidth, or 2400 b/s for a channel having 1200 Hz analog bandwidth. The bottom sketch shows that if each half-cycle conveys one of sixteen amplitude levels, providing the channel has sufficiently low noise, four bits can be coded per half-cycle. The rate at which the signal in the channel can change state – the *symbol rate* or *baud rate* – is constant at 2400 baud, but this modulation method has a *data rate* or *bit rate* of 9600 b/s.

Resolution

As picture detail increases in frequency, the response of an imaging system will eventually deteriorate. In image science and in television, *resolution* refers to the capability of an imaging system to reproduce fine detail in the picture.

The absolute upper limit to resolution in a digital image system is the number of pixels over the width and height of a frame, and is the way the term *resolution* is used in computing.

In conventional North American television, 483 scan lines cover the height of the image. High-definition television systems use up to 1080 picture lines. The amount of information that can be captured in a video signal is bounded by the number of picture lines. But other factors impose limits more severe than the number of lines per picture height.

In an interlaced system, vertical resolution must be reduced substantially from the scan-line limit, in order to avoid producing a signal that will exhibit objectionable twitter upon display.

Resolution in film

In film, resolution is measured as the finest pattern of straight, parallel lines that can be reproduced, expressed in *line pairs per millimeter* (lp/mm). A line pair contains a black region and a white region.

Motion picture film is conveyed vertically through the camera and projector, so the width – not the height – of the film is 35 mm. Cinema usually has an aspect ratio of 1.85:1, so the projected film area is about 21 mm × 11 mm, only three-tenths of the 36 mm × 24 mm projected area of 35 mm still film.

The limit to the resolution of motion picture film is not the static response of the film, but judder and weave in the camera and the projector.

Resolution in television

In video, resolution refers to the number of line pairs (cycles) resolved on the face of the display screen, expressed in cycles per picture height (C/PH) or cycles per picture width (C/PW). A *cycle* is equivalent to a *line pair* of film. In a digital system, it takes at least two samples – pixels, scanning lines, or *TV lines* – to represent a line pair. However, resolution may be substantially less than the number of pixel pairs due to optical, electro-optical, and electrical filtering effects. *Limiting resolution* is defined as the frequency where detail is recorded with just 10 percent of the system's low-frequency response.

In consumer television, the number of scanning lines is fixed by the raster standard, but the electronics of transmission, recording, and display systems tend to limit bandwidth and reduce horizontal resolution. Consequently, in consumer electronics the term *resolution* generally refers to horizontal resolution. Confusingly, horizontal resolution is expressed in units of lines per picture *height*, so once the number of resolvable lines is measured, it must be corrected for the aspect ratio of the picture. Resolution in *TV lines per picture height* is twice the resolution in cycles per picture width, divided by the aspect ratio of the picture.

Resolution in computer graphics

In computer graphics, *resolution* is simply the number of discrete vertical and horizontal pixels required to store the digitized image. For example, a 1152×900 system has a total of about one million pixels (one megapixel, or 1 Mpx). Computer graphics is not generally very concerned about whether individual pixels can be discerned on the face of the display. In most color computer systems, an image comprising a one-pixel black-and-white checkerboard actually displays as a uniform gray, due to poor high-frequency response in the cable and video amplifiers, and due to rather large spot size at the CRT.

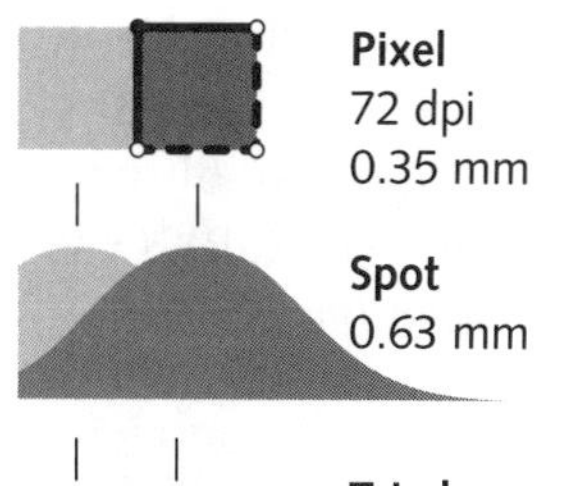

Figure 1.13
Pixel/spot/triad.

Computer graphics often treats each pixel as representing an idealized rectangular area independent of all other pixels. This notion discounts the correlation among pixels that is an inherent and necessary aspect of image acquisition, processing, compression, display, and perception. In fact the rather large spot produced by the electron beam of a CRT and the arrangement of phosphor triads on the screen, suggested by Figure 1.13, produces an image of a pixel on the screen that bears little resemblance to a rectangle. If pixels are viewed at a sufficient distance, these artifacts are of little importance. However, imaging systems are forced by economic pressures to make maximum perceptual use of the delivered pixels, consequently we tend to view CRTs at close viewing distances.

Luma

As you will see in *Luma and color differences*, on page 155, a video system conveys image data in the form of a component that represents brightness, and two other components that represent color. It is important to convey the brightness component in such a way that noise (or quantization) introduced in transmission, processing, and storage has a perceptually similar effect across the entire tone scale from black to white. Ideally, these goals would be accomplished by forming a true CIE luminance signal as a weighted sum of linear-light red, green, and blue; then subjecting that luminance to a nonlinear transfer function similar to the CIE L^* function that will be described on page 88.

There are practical reasons in video to perform these operations in the opposite order. First a nonlinear transfer function – *gamma correction* – is applied to each of the linear R, G, and B. Then a weighted sum of the nonlinear components is computed to form a *luma* signal, Y', representative of brightness.

625/50 standards documents indicate a precorrection of $1/_{2.8}$, approximately 0.36, but this value is rarely used in practice. See *Gamma* on page 91.

In effect, video systems approximate the lightness response of vision using RGB intensity signals, each raised to the 0.45 power. This is comparable to the $1/_3$ power function defined by L^*.

Recommendation ITU-R BT.601-4, *Encoding Parameters of Digital Television for Studios.* Geneva: ITU, 1990.

The coefficients that correspond to the so-called *NTSC* red, green, and blue CRT phosphors of 1953 are standardized in Recommendation ITU-R BT. 601-4 of the ITU Radiocommunication Sector (formerly CCIR). I call it *Rec. 601*. To compute nonlinear video *luma* from nonlinear red, green, and blue:

Eq 1.2

$$Y'_{601} = 0.299\,R' + 0.587\,G' + 0.114\,B'$$

The prime symbols in this equation, and in those to follow, denote nonlinear components.

The unfortunate term "video luminance"

Unfortunately, in video practice, the term *luminance* has come to mean *the video signal representative of luminance* even though the components of this signal have been subjected to a nonlinear transfer function. At the dawn of video, the nonlinear signal was denoted *Y'*, where the prime symbol indicated the nonlinear treatment. But over the last 40 years the prime has been elided and now both the term *luminance* and the symbol *Y* collide with the CIE, making both ambiguous! This has led to great confusion, such as the incorrect statement commonly found in computer graphics and color textbooks that in the *YIQ* or *YUV* color spaces, the *Y* component *is* CIE luminance! I use the term *luminance* according to its standardized CIE definition and use the term *luma* to refer to the video signal, and I am careful to designate the nonlinear quantity with a prime symbol. But my convention is not yet widespread, and in the meantime you must be careful to determine whether a linear or nonlinear interpretation is being applied to the word and the symbol.

Color difference coding

In component video, the three components necessary to convey color information are transmitted separately.

The data capacity accorded to the color information in a video signal can be reduced by taking advantage of the relatively poor color acuity of vision, providing full

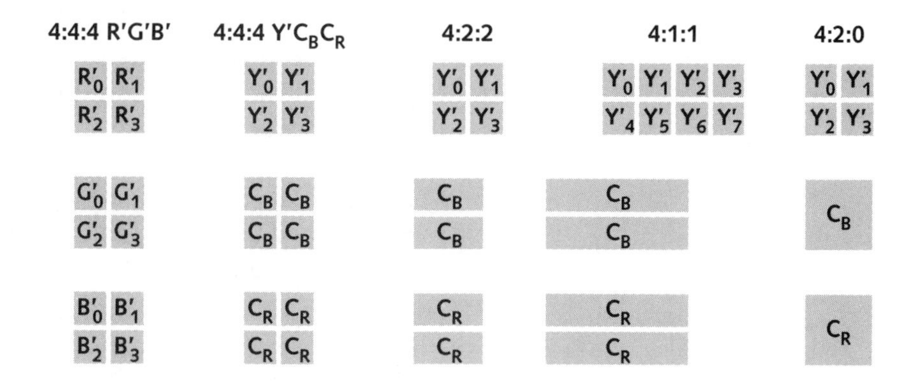

Figure 1.14 **Chroma subsampling.** A 2×2 array of *R'G'B'* pixels can be transformed to a luma component *Y'* and two color difference components C_B and C_R; color detail can then be reduced by subsampling, provided that full luma detail is maintained. The wide aspect of the C_B and C_R samples indicates their spatial extent. The horizontal offset of C_B and C_R in 4:2:2 and 4:1:1 sampling is due to cositing.

luma bandwidth is maintained. It is ubiquitous to base *color difference* signals on *blue minus luma* and *red minus luma* (*B'*-*Y'*, *R'*-*Y'*). Luma and (*B'*-*Y'*, *R'*-*Y'*) can be computed from *R'*, *G'*, and *B'* through a 3×3 matrix multiplication. Once luma and color difference – or *chroma* – components have been formed, the chroma components can be subsampled (filtered).

$Y'C_BC_R$

In component digital video, C_B and C_R components scaled from (*B'*-*Y'*, *R'*-*Y'*) are formed.

$Y'P_BP_R$

In component analog video, P_B and P_R color difference signals scaled from (*B'*-*Y'*, *R'*-*Y'*) are lowpass filtered to about half the bandwidth of luma.

4:4:4

In Figure 1.14 above, the left-hand column sketches a 2×2 array of *R'G'B'* pixels that, with 8 bits per sample, would occupy a total of 12 bytes. This is denoted 4:4:4 *R'G'B'*. $Y'C_BC_R$ components can be formed from *R'G'B'*, as shown in the second column; without subsampling, this is denoted 4:4:4 $Y'C_BC_R$.

The use of *4* as the numerical basis for subsampling notation is a historical reference to a sample rate of about four times the color subcarrier frequency.

4:2:2

$Y'C_BC_R$ digital video according to Rec. 601 uses 4:2:2 sampling: Chroma components are subsampled by a factor of 2 along the horizontal axis. Chroma samples are coincident (cosited) with alternate luma samples.

In an 8-bit system using 4:2:2 coding, the 2×2 array occupies 8 bytes, and the aggregate data capacity is 16 bits per pixel. For studio digital video, the raw data rate is 27 MB/s.

4:1:1

A few digital video systems have used 4:1:1 sampling, where the chroma components are subsampled by a factor of 4 horizontally.

4:2:0

In 4:2:0 sampling, usually used in JPEG and MPEG, C_B and C_R are each subsampled by a factor of 2 in both the horizontal and vertical axes. Chroma samples are sited *interstitially*, vertically halfway between lines, and, unlike the cositing of 4:2:2, are sited horizontally halfway between alternate luma samples. In 8-bit $Y'C_BC_R$ coded 4:2:0 or 4:1:1, the aggregate uncompressed data capacity is 12 bits per pixel.

MAC

A transmission system for analog components – *Multiplexed Analog Components*, or MAC – has been adopted in Europe for direct broadcast from satellite (DBS). In MAC, the color difference components are not combined with each other or with luma, but are time-compressed and transmitted serially. MAC is not standardized by ITU-R.

Component digital video, 4:2:2

The standard interface for 4:2:2 component digital video is Rec. ITU-R 601-4. It specifies sampling of luma at 13.5 MHz and sampling of C_B and C_R color difference components at 6.75 MHz. This interface is referred to as *4:2:2*, since luma is sampled at four times 3.375 MHz, and each of the C_B and C_R components at twice 3.375 MHz – that is, the color difference signals are horizontally subsampled by a factor of 2:1 with respect to luma. Sampling at 13.5 MHz results in an integer number of *samples per total line* (S/TL) in both

A version of Rec. 601 uses 18 MHz sampling to produce a picture aspect ratio of 16:9.

525/59.94 systems (858 S/TL) and 625/50 systems (864 S/TL). Luma is sampled with 720 *active* samples per line in both 525/59.94 and 625/50.

Component digital video tape recorders are widely available for both 525/59.94 and 625/50 systems, and have been standardized with the designation *D-1*. That designation properly applies to the tape format, not the signal interface.

Transport, electrical, and mechanical aspects of 4:2:2 interface are specified in Rec. 656. See page 248.

Rec. 601 specifies luma coding that places black at code 16 and white at code 235. Color differences are coded in offset binary, with zero at code 128, the negative peak at code 16, and the positive peak at code 240.

Composite video

The terms *NTSC* and *PAL* are often used incorrectly to refer to scanning standards. Since PAL encoding is used with both 625/50 scanning (with two different subcarrier frequencies) and 525/59.94 scanning (with a third subcarrier frequency), the term *PAL* alone is ambiguous. The notation *CCIR* is sometimes used to refer to 625/50 scanning, but that is confusing because the former CCIR – now ITU-R – standardized all scanning systems, not just 625/50.

In composite NTSC and PAL video, the color difference signals required to convey color information are combined by the technique of quadrature modulation into a *chroma* signal using a color subcarrier of about 3.58 MHz in conventional NTSC and about 4.43 MHz in conventional PAL. Luma and chroma are then summed into a composite signal for processing, recording, or transmission. Summing combines brightness and color into one signal, at the expense of introducing a certain degree of mutual interference.

The frequency and phase of the subcarrier are chosen and maintained carefully: The subcarrier frequency is chosen so that luma and chroma, when they are summed, are *frequency interleaved*. Studio signals have coherent sync and color subcarrier; that is, subcarrier is phase-locked to a rational fraction of the line rate; generally this is achieved by dividing both from a single master clock. In industrial and consumer video, subcarrier usually free-runs with respect to line sync.

SECAM sums luma and chroma without using frequency interleaving. SECAM has no application in the studio. See page 254.

Transcoding among different color encoding methods having the same raster standard is accomplished by luma/chroma separation, color demodulation, and color remodulation.

Composite digital video, $4f_{SC}$

The earliest digital video equipment processed signals in composite form. Processing of digital composite signals is simplified if the sampling frequency is an integer multiple of the color subcarrier frequency. Nowadays, a multiple of four is used: *four-times-subcarrier*, or $4f_{SC}$. For NTSC systems it is standard to sample at about 14.3 MHz. For PAL systems the sampling frequency is about 17.7 MHz.

Composite digital processing was necessary in the early days of digital video, but most image manipulation operations cannot be accomplished in the composite domain. During the 1980s there was widespread deployment of component digital processing equipment and component videotape recorders (DVTRs), recording 4:2:2 signals using the D-1 standard.

However, the data rate of a component 4:2:2 signal is roughly twice that of a composite signal. Four-times-subcarrier composite digital coding was resurrected to enable a cheap DVTR; this became the *D-2* standard. The user of a D-2 DVTR has the advantages of digital recording, but retains the disadvantages of composite NTSC or PAL: Luma and chroma are subject to cross-contamination, and the pictures cannot be manipulated without decoding and reencoding.

The development and standardization of D-2 recording led to the standardization of composite $4f_{SC}$ digital parallel and serial interfaces, which essentially just code the raw 8- or 10-bit composite data stream. These interfaces share the electrical and physical characteristics of the standard 4:2:2 interface, but with about half the data rate. For 8-bit sampling this leads to a total data rate of about 14.3 MB/s for 525/59.94 NTSC, and about 17.7 MB/s for 625/50 PAL.

Analog interface

Video signal amplitude levels in 525/59.94 systems are expressed in IRE units, named after the Institute of Radio Engineers in the United States, the predecessor

In 525/59.94 with setup, *picture excursion* refers to the range from blanking to white, even though strictly speaking the lowest level of the picture signal is 7.5 IRE, and not 0 IRE.

of the IEEE. Reference blanking level is defined as 0 IRE, and reference white level is 100 IRE. The range between these values is the *picture excursion*.

Composite 525/59.94 systems have a picture-to-sync ratio of 10:4; consequently, the sync level of a composite 525/59.94 signal is -40 IRE. In composite 525/59.94 systems, reference black is *setup* the fraction 7.5 percent ($3/40$) of the reference blanking-to-white excursion: Composite 525/59.94 employs a *pedestal* of 7.5 IRE. There are exactly 92.5 IRE from black to white: The picture excursion of a 525/59.94 signal is about 661 mV.

Setup has been abolished from component digital video and from HDTV. Many 525/59.94 component analog systems have adopted *zero setup*, and have 700 mV excursion from black to white, with 300 mV sync. But many component analog systems use setup, and it is a nuisance in design and in operation.

625/50 systems have a picture-to-sync ratio of 7:3, and zero setup. Picture excursion (from black to white) is exactly 700 mV; sync amplitude is exactly 300 mV. Because the reference levels are exact in millivolts, the IRE unit is little used, but in 625/50 systems an IRE unit corresponds to exactly 7 mV.

A video signal with sync is distributed in the studio with blanking level at zero (0 V_{DC}) and an amplitude from synctip to reference white of one volt into an impedance of 75 Ω. A video signal without sync is distributed with blanking level at zero, and an amplitude from synctip to reference white of either 700 mV or 714 mV.

High-definition television (HDTV)

High-definition television (HDTV) is defined as having twice the vertical and twice the horizontal resolution of conventional television, a picture aspect ratio of 16:9, a frame rate of 24 Hz or higher, and at least two channels of CD-quality sound.

HDTV studio equipment is commercially with 1125/60/2:1 scanning and 1920×1035 image format, with about two megapixels per frame – six times the number of pixels of conventional television. The data rate of studio-quality HDTV is about 120 megabytes per second. Commercially available HDTV cameras rival the picture quality of the best motion picture cameras and films.

NHK Science and Technical Research Laboratories, *High Definition Television: Hi-Vision Technology.* New York: Van Nostrand Reinhold, 1993.

SMPTE 274M-1995, *1920×1080 Scanning and Interface.*

Except for their higher sampling rates, studio standards for HDTV have a close relationship to studio standards for conventional video, which I will describe in the rest of the book. For details specific to HDTV, consult the book from NHK Labs, and SMPTE 274M.

Advanced Television (ATV) refers to transmission systems designed for the delivery of entertainment to consumers, at quality levels substantially improved over conventional television. ATV transmission systems based on 1125/60/2:1 scanning and MUSE compression have been deployed in Japan. The United States is poised to adopt standards for ATV based on 1920×1080 and 1280×720 image formats. MPEG-2 compression can compress this to about 20 megabits per second, a rate suitable for transmission through a 6 MHz terrestrial VHF/UHF channel.

The compression and digital transmission technology developed for ATV has been adapted for digital transmission of conventional television; this is known as *standard-definition television* (SDTV). MPEG-2 compression and digital transmission allow a broadcaster to place about four digital channels in the bandwidth occupied by a single analog NTSC signal. Digital television services are already deployed in direct broadcast satellite (DBS) systems and are expected soon in cable television (CATV).

The MUSE system adopted in Japan is documented in the book from NHK Labs. Digital ATV broadcasting standards are not yet formally adopted.

With the advent of HDTV, 16:9 widescreen variants of conventional 525/59.94 and 625/50 component video have been proposed and even standardized. In studio analog systems, widescreen is accomplished by having the active picture represent 16:9 aspect ratio, but keeping all of the other parameters of the video standards. Unless bandwidth is increased by the same $\frac{4}{3}$ ratio as the increase in aspect ratio, horizontal detail suffers.

In digital video, there are two approaches to achieving 16:9 aspect ratio. The first approach is comparable to the analog approach that I mentioned a moment ago: The sampling rate remains the same as conventional component digital video, and horizontal resolution is reduced by a factor of $\frac{3}{4}$. In the second approach, the sampling rate is increased from 13.5 MHz to 18 MHz. I consider all of these schemes to adapt conventional video to widescreen be unfortunate: None of them offers an increase in resolution sufficient to achieve the product differentiation that is vital to the success of any new consumer product.

Raster images in computing 2

Video engineering was once a highly specialized domain. But the application of digital technology in diverse fields now means that video can be considered as a subset of the larger class of digital imaging: Video involves raster image data in time sequence. This chapter places video in the context of other digital imaging technologies, particularly the technologies of computer graphics, as shown in Figure 2.1 below.

Figure 2.1 **Raster images in computing.**

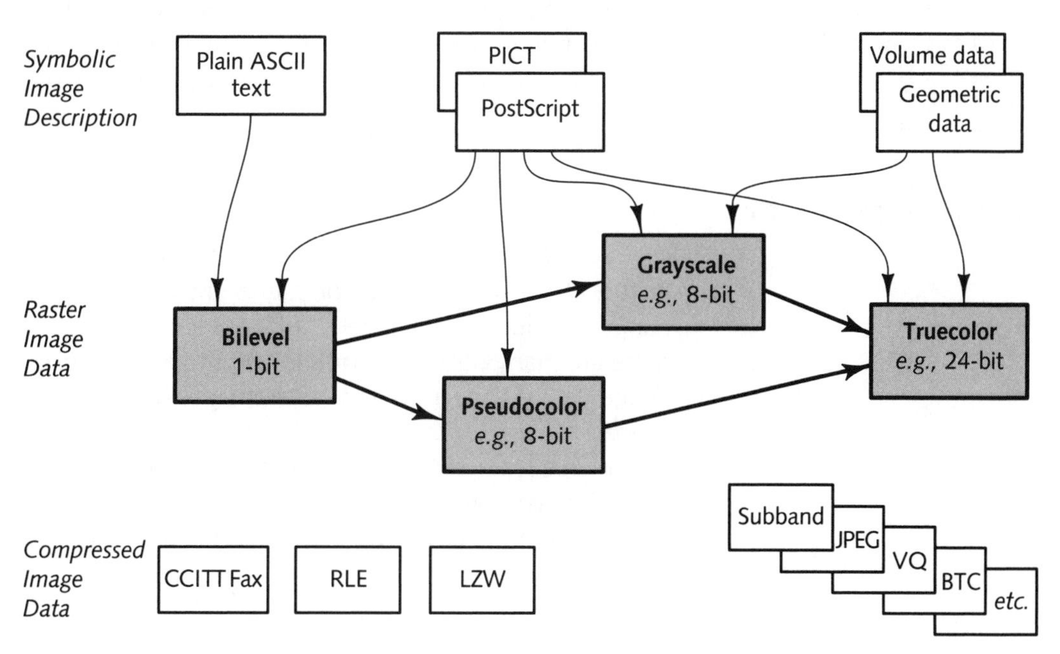

Introduction

Figure 2.1 has three rows, corresponding to three basic representations of an image: *symbolic image description, raster image,* and *compressed image*.

- **Symbolic image description** A description of the image in symbolic form. Not an image itself, but a high-level description of an image.

- **Raster image** A direct representation of a sampled image, enumerating the grayscale or color content of each pixel in scan-line order. There are four fundamental types of raster image: *bilevel*, *pseudocolor, grayscale,* and *truecolor.* In Figure 2.1, I have arranged these in four columns to show successively higher quality levels, from low quality at the left to high quality at the right. Video properly concerns only grayscale and truecolor images.

- **Compressed image** A raster image that has been compressed by a method that reduces storage or transmission data requirement. Several compressed image formats are shown in the bottom row.

James D. Murray and William vanRyper, *Encyclopedia of Graphics File Formats.* Sebastopol, CA: O'Reilly & Associates, 1994.

Many different file formats are in use for each of these representations. Discussion of file formats *per se* is outside the scope of this book. To convey high-quality color images, a file format must accommodate at least 24 bits per pixel.

Symbolic image description

Many methods are in use to describe the content of a picture at a level of abstraction higher than enumerating the value of each pixel. PostScript is common; there are many other less common methods. Symbolic image data is converted to a raster image by the process of *rasterizing*. ASCII text can be considered to be a highly restricted symbolic description, where rasterizing comprises imaging the graphic representations – or *glyphs* – corresponding to each of the codes.

Images are *rasterized* – or *imaged* or *rendered* – by interpreting symbolic data and producing raster image data. In the roadmap, this operation passes information from the top row to the middle row.

Raster image data cannot easily be transformed back to its symbolic description form once rasterized: A raster image – in the middle row of the roadmap – cannot be returned to its description in the top row. If your application involves rendered images, you may find it useful to retain the symbolic data even after rendering, in case the need arises to rerender the image – at a different size, perhaps – or to perform a modification, such as removing an object.

Images from a fax machine, a grayscale or color scanner, or a video camera originate in raster image form: No symbolic description is available. Optical character recognition (OCR) and raster-to-vector techniques make brave but generally unsatisfying attempts to extract text or geometric data from raster images.

In PostScript, an image is described in the form of a program in a language specialized for imaging operations. When the program is executed by a PostScript interpreter, the image is drawn. In PostScript, the rasterizing operation is often called *raster image processing*, or *ripping*.

Geometric data describes the properties and positions of objects. Geometric data may be interpreted to produce an image from a particular viewpoint. Rasterizing from geometric data is called *rendering;* usually, truecolor images are produced.

Raster images

There are four distinct types of raster image data: *bilevel* (by definition, 1 bit per pixel), *pseudocolor* (typically 8 bits per pixel), *grayscale* (typically 8 bits per pixel), and *truecolor* (typically 24 bits per pixel).

Bilevel

Each pixel of a bilevel image has a single bit that represents either black or white – but nothing in between. In computing this is often called *monochrome*, but that term properly refers to an image containing shades of a single color and therefore should apply to both bilevel and grayscale images.

Since the first days of data communications, binary zero has been known as *space,* and binary one has been known as *mark*. In video, a "mark" on a CRT emits light, so in video it is conventional for binary one (or the maximum code value) to represent white. In printing, a "mark" deposits ink on the page. It is conventional in printing for a binary one (or the maximum code value) to represent black.

Pseudocolor

In a *pseudocolor, indexed color,* or *colormapped* system, several bits – often 8 – are dedicated to each pixel in an image or framebuffer. This provides a moderate number of unique codes, often 256, to each pixel. The code of each pixel is used as an index into a *color lookup table* (CLUT, or *colormap* or *palette*) that retrieves red, green, and blue components associated with the code. The CLUT may have 4, 6, 8, or more bits for each of the red, green, and blue components.

In an 8-bit pseudocolor system, any particular image – or the content of the framebuffer at any instant in time – is limited to a selection of just 256 colors from the universe of available colors. A typical lookup table has 8-bit values for each of red, green, and blue, so each pixel can be chosen from a large universe of 2^{24} or 16777216 colors. A pseudocolor image can be thought of as "painting by numbers," where the number of colors is rather small.

In pseudocolor, the colors assigned to adjacent codes are generally completely unrelated: The color assigned to code 42 has no proximity to the color assigned to code 43. Pseudocolor is appropriate for images such as maps, schematic diagrams, or cartoons, where several different ink colors and combinations are used but where each color or combination is either completely present or completely absent at any point in the image.

Grayscale

A grayscale image represents a continuous range of tones from black, through many intermediate shades of gray, to white. A grayscale system can represent a black-and-white photograph. In printing, a grayscale image is called continuous tone (*contone*), as distinct from *line art* or *type*.

Truecolor

A truecolor system has separate red, green, and blue components for each pixel of an image or in a framebuffer. In most computer systems, each component is represented by a *byte* of 8 bits, so each pixel has 24 bits of color information.

In a straightforward design for a framebuffer memory system, each pixel has a number of bytes that is a power of 2. A truecolor framebuffer often has 4 bytes per pixel. Three bytes are used for red, green, and blue color components; the fourth byte is used for purposes other than representing color. The fourth byte may contain overlay information. Alternatively, the fourth byte may store an *alpha* component representing opacity from zero (fully transparent) to unity (fully opaque). In computer graphics, the *alpha* component is conventionally in the intensity (linear-light) domain. In video, this component is called *linear key*, but it is not linear in *intensity* – instead, it is linear in gamma-corrected *voltage*.

A truecolor framebuffer usually has three lookup tables (LUTs), one for each component. In X Window System terminology, *truecolor* implies that the lookup tables can be altered under control of the application program. If a truecolor framebuffer has no output lookup tables, or LUTs that cannot be altered under program control, in X it is said to have *direct color*.

The *RGB* components of each pixel in a 24-bit system can represent 16.7 million distinct codes, but the number of these codes that represent colors that can be distinguished depends on the transfer function used. With 24-bit color and properly chosen transfer functions, near-photographic quality images can be displayed and geometric objects can be smoothly shaded. But in a linear-light representation with 8 bits per component, contouring will be evident in many scenes: Having 24-bit color is not a guarantee of good image quality.

Dithering

A display device may have only a small number of grayscale values or color values at each device pixel. However, if the viewer is sufficiently distant from the display, the value of neighboring pixels can be set so that the viewer's eye integrates several pixels to achieve an apparent improvement in the number of levels or colors that can be reproduced.

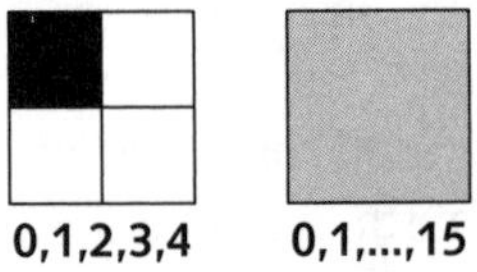

0,1,2,3,4 **0,1,...,15**

Figure 2.2 **Dithering.**

Suppose each pixel in the 2×2 array in Figure 2.2 in the margin can be set either black or white. Four bits are required to control the pixels, and 16 patterns can be created, but only five grayscale shades can be produced: $^0/_4$, $^1/_4$, $^2/_4$, $^3/_4$, and $^4/_4$. If instead four binary-coded bits are used to control a single grayscale pixel occupying the same area, 16 shades can be achieved: Grayscale representation is more memory efficient than dithering.

A computer display is normally viewed from a distance where a device pixel subtends a rather large angle at the viewer's eye, relative to his spatial acuity. Applying dither to a conventional computer display often introduces objectionable artifacts. However, careful application of dither can be effective. For example, human vision has poor acuity for blue spatial detail but good color discrimination capability in blue. Blue can be dithered across 2×2 pixel arrays to produce four times the number of apparent blue levels, with no perceptible penalty at normal viewing distances.

Dithering techniques can be used for print reproduction, but sophisticated print systems use *halftoning*, which will be described on page 110.

Conversion among types

Bilevel raster image data can be *widened* to any other form easily, simply by assigning the codes for black and white. A grayscale image can be converted to truecolor by assigning codes from black to white. A pseudocolor image can be widened into truecolor by application – usually in software – of the color lookup table. These transformations correspond to left-to-right traversal of the middle row of Figure 2.1. Widening adds bits but not information. Nonetheless it is useful for displaying a bilevel image in a pseudocolor or grayscale system, or displaying a bilevel, pseudocolor, or grayscale image in a truecolor display system.

I constructed Figure 2.1 so that traversal of the diagram from left to right, and from top to bottom, corresponds to conversions that can be accomplished without loss. Traversal from right to left in the diagram, or bottom to top, generally involves loss. However, if a pseudocolor, grayscale, or truecolor image contains only the two colors black and white, it can be *narrowed* to bilevel without loss.

Truecolor image data can be approximated in pseudocolor form either through the use of a fixed colormap, or through a *colormap quantization* algorithm that examines the truecolor image and computes a colormap that is in some sense best for a particular image or sequence of images.

Pseudocolor and grayscale images are fundamentally at odds, except that a pseudocolor system may be operated in grayscale mode if the loss of color can be tolerated. A pseudocolor system can accommodate grayscale images by having its colormap divided into a grayscale section and a color section. The penalty is a compromise in quality for both types of images – a loss of available color choices and a loss of grayscale codes.

The grayscale component of a truecolor image can be computed relatively easily, but this capability is generally important only for reverse compatibility with grayscale displays, as in watching a color program on a black-and-white television set.

Data compression

Data compression reduces the number of bits required to store or convey text, numeric, binary, image, sound, or other data by exploiting statistical properties of the data. Data compression attempts to reduce the number of bits needed to store or transmit data, at the expense of some computational effort to compress and decompress.

Image compression

Binary data typical of general computer applications often has patterns of repeating byte strings and substrings. Data compression techniques, such as *run-length encoding* (RLE) and *Lempel-Ziv-Welch* (LZW), take advantage of repeated information in order to accomplish compression. When applied to bilevel or pseudocolor image data stored in scan-line order, the RLE and LZW algorithms exploit some amount of the horizontal correlation, so RLE and LZW compress these images somewhat, perhaps 3:1.

Digital image data typically has strong vertical and horizontal correlations among pixels. If a data compression algorithm is designed to exploit the statistics of image data, as opposed to arbitrary binary data, then better compression is possible. For example, the ITU-T (former CCITT) fax standard for bilevel image data exploits correlation in vertical and horizontal directions, and achieves compression much better than RLE or LZW. But the algorithm is designed for bilevel images, and would perform poorly if it were applied to arbitrary binary data.

Lossy compression

Data compression methods reproduce exactly, bit for bit, the data presented to the compressor: Data compression is by definition *lossless*. Image and sound data is generally quite redundant, and lossless techniques optimized for images or sounds can result in reasonably good compression. But the characteristics of human perception can be exploited to achieve much higher compression ratios if the requirement to exactly reproduce the input bit stream is relaxed. Image or sound data can be subject to *lossy* compression, where no guarantee of bit for bit reconstruction is made, provided that the impairments introduced are not overly perceptible. Approximations can be made that dramatically improve the effectiveness of the compression, while causing minimal perceptual impairment.

John Watkinson, *Compression in Video & Audio*. Sevenoaks, Kent, England: Focal Press, 1995.

Majid Rabbani and Paul W. Jones, *Digital Image Compression Techniques*. Bellingham, WA: SPIE Optical Engineering Press, 1991.

Lossy compression schemes are almost never used for bilevel or pseudocolor images. Lossy compression is very effective for continuous tone (grayscale or truecolor) images.

Filtering and sampling 3

This chapter explains how a one-dimensional signal is filtered and sampled, and how it is reconstructed following D-to-A conversion. In the following chapter, *Image digitization and reconstruction*, on page 67, I extend these concepts to the two dimensions of an image and the three dimensions of video.

Introduction

Sampling theory was originally developed to describe one-dimensional signals such as audio, where the signal is a continuous function of the single dimension of time. Sampling theory has been extended to images, where an image is treated as a continuous function of two spatial coordinates (horizontal and vertical). Sampling theory can be applied to moving images, where the third coordinate is time. This chapter will explain the concepts of sampling in the one-dimensional case, as in sampling an audio signal. The next chapter will extend the explanation to images.

My explanation describes the original sampling of an analog signal waveform. If you are more comfortable remaining in the digital domain, consider the problem of shrinking a scan line of image samples by a factor of n (say, $n = 16$) to accomplish image resizing. You need to compute one output sample for each set of n input samples. This is the *resampling* problem in the digital domain. Its constraints are virtually identical to the constraints of original sampling of an analog signal.

When audio is sampled, each sample must encapsulate the potentially complex waveform during the sample period. When an image is sampled, each sample encapsulates a potentially complex distribution of power over a small region of the image plane. In each case, a potentially vast amount of information is reduced to a single number.

Upon sampling, detail within the sample interval must be discarded. The challenge of sampling is to determine ways to discard this information, while avoiding the loss of information at scales larger than the sample pitch, all the time avoiding the introduction of artifacts. *Sampling theory* elaborates the conditions under which a signal can be sampled and accurately reconstructed, subject only to inevitable loss of detail.

Sampling theorem

Assume that a signal to digitized is well behaved, changing relatively slowly as a function of time. Consider the cosine signals shown in Figure 3.1 below, where the x-axis shows sample intervals. The top waveform is a cosine at the fraction 0.35 of the sampling rate f_S; the middle waveform is at $0.65 f_S$. The bottom row shows that identical samples result from sampling either of these waveforms: Either of the waveforms can masquerade as the same sample sequence. If the middle waveform is sampled, then reconstructed conventionally, the top waveform will result. This is the phenomenon of *aliasing*.

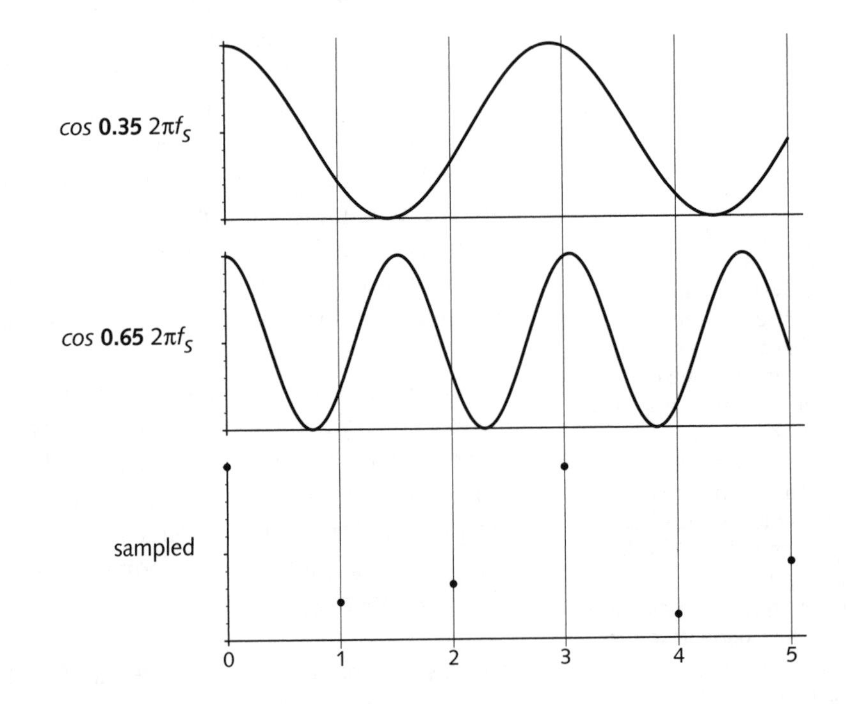

Figure 3.1 **Cosine waves less than and greater than $0.5 f_S$.** These two cosine waves – at the fractions 0.35 and 0.65 of the sampling rate – produce exactly the same set of sampled values when point-sampled.

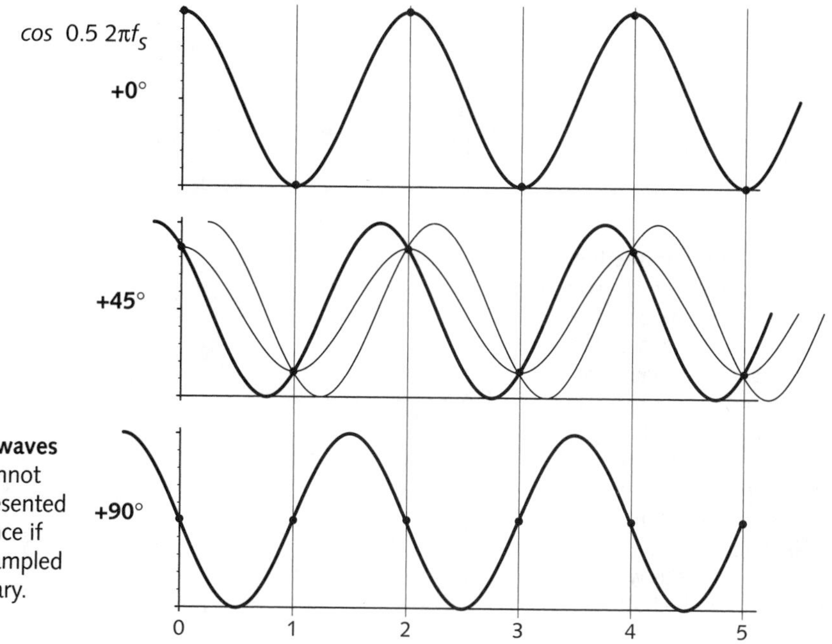

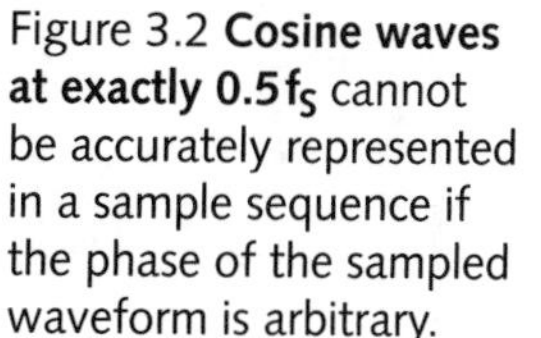

Figure 3.2 **Cosine waves at exactly $0.5 f_S$** cannot be accurately represented in a sample sequence if the phase of the sampled waveform is arbitrary.

Sampling at exactly $0.5f_S$

You might assume that a signal whose frequency is exactly half the sampling rate can be accurately represented by an alternating sequence of sample values, say, zero and one. In Figure 3.2 above, the series of samples in the top row is unambiguous (provided it is known that the amplitude of the waveform is unity). But the samples of the middle row could be generated from any of the three indicated waveforms, and the phase-shifted waveform in the bottom row has samples that are indistinguishable from a constant waveform. The inability to accurately analyze a signal at exactly half the sampling frequency leads to the strict "less-than" condition in Shannon's theorem, which I will now describe.

Shannon in the United States, and Kotelnikov in Russia, independently concluded that to guarantee sampling of a signal without the introduction of aliases, all of the signal's frequency components must be contained within half the sampling rate (now

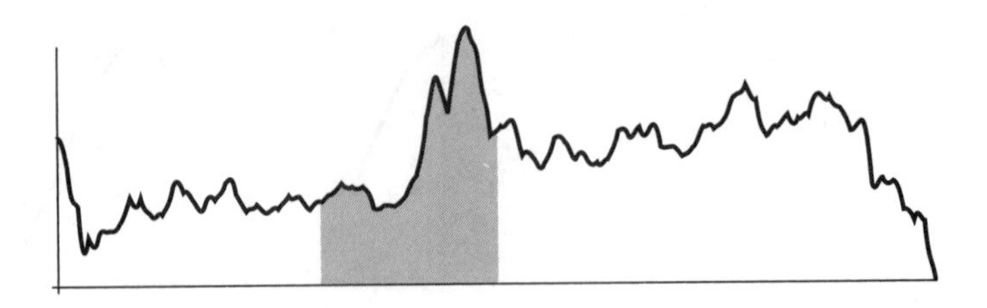

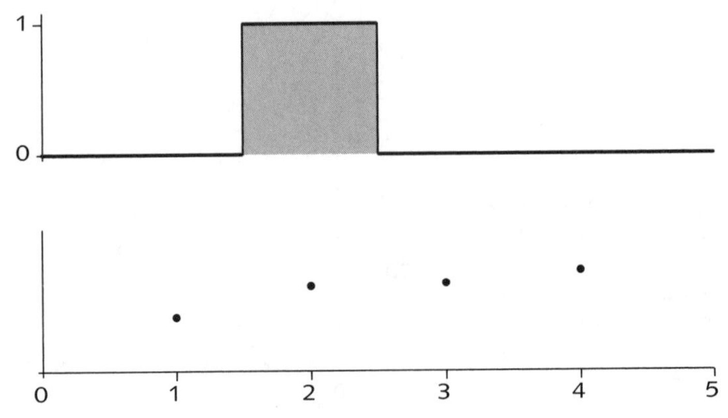

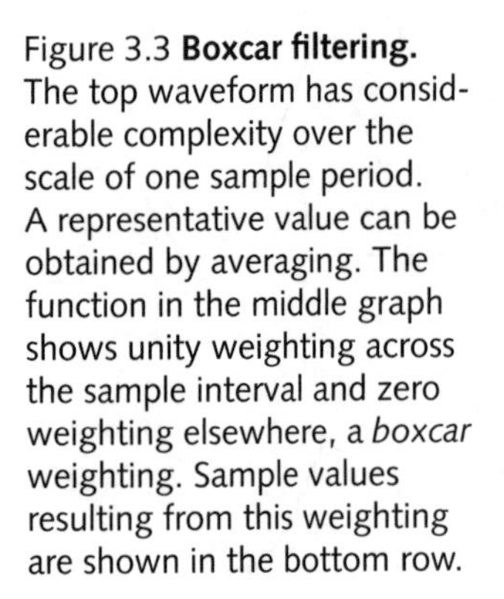

Figure 3.3 **Boxcar filtering.** The top waveform has considerable complexity over the scale of one sample period. A representative value can be obtained by averaging. The function in the middle graph shows unity weighting across the sample interval and zero weighting elsewhere, a *boxcar* weighting. Sample values resulting from this weighting are shown in the bottom row.

known as the *Nyquist frequency*). The condition is usually imposed by analog filtering prior to sampling, to remove frequency components at $0.5f_S$ and higher. A filter must implement some sort of integration. In the example of Figure 3.1, no filtering was performed; the waveform was simply *point-sampled*. The lack of filtering admitted aliases.

Perhaps the most basic way to filter the waveform across each sample period is to average the waveform for that duration. The top graph of Figure 3.3 above illustrates this averaging: The shading under the waveform indicates the integration operation. Many different integration schemes are possible; these can be represented as weighting functions plotted as a function of time. Simple averaging uses a weighting function whose value is unity during the sample period and zero outside that interval, as sketched in the middle graph of Figure 3.3. This operation is called *boxcar* filtering, since a sequence of these functions with different amplitudes resembles the profile of a freight

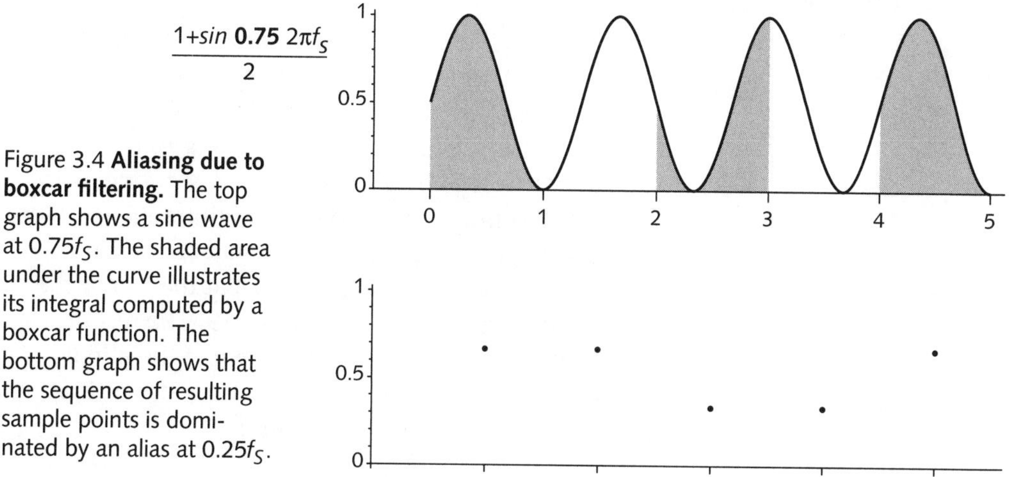

Figure 3.4 **Aliasing due to boxcar filtering.** The top graph shows a sine wave at $0.75f_S$. The shaded area under the curve illustrates its integral computed by a boxcar function. The bottom graph shows that the sequence of resulting sample points is dominated by an alias at $0.25f_S$.

train. Once sampled, the signal is represented by discrete values. I plot these as points in the bottom row of the figure. To plot these values as amplitudes of the boxcar function would wrongly suggest that a boxcar function should be used as a reconstruction filter.

A problem with boxcar filtering is evident in Figure 3.4 above. The top graph shows a sine wave at $0.75f_S$; it exceeds the Nyquist frequency. The shaded regions show integration over intervals of one sample period. For the sine wave at $0.75f_S$, sampled starting at zero phase, the first pair of integrated values are ($2/3$, $2/3$) and the second pair are ($1/3$, $1/3$). The dominant component of the filtered sample sequence, shown in the bottom graph, is one-quarter of the sampling frequency. Filtering using a boxcar weighting function has created an unwanted alias.

Figure 3.4 is another example of *aliasing*: Due to a poor presampling filter, the sequence of sampled values exhibits a frequency component not present in the input signal. Boxcar integration is not sufficient to prevent fairly serious aliasing in this example.

To gain a general appreciation of aliasing, it is necessary to understand signals in the *frequency domain*.

Frequency response

The previous section gave an example of filtering prior to sampling that produced an unexpected alias. You can determine whether a filter has an unexpected response at *any* frequency by presenting to the filter a signal that sweeps through all frequencies, from zero, through low frequencies, to some high frequency, plotting the amplitude response of the filter as you go. I sketched a frequency sweep signal at the top of the graph *Frequency response* on page 19. The response waveform of a lowpass filter is shown in the middle graph of that figure; the amplitude response of that filter is shown in the bottom graph.

Amplitude response is the maximum response over all phases of the input signal at each frequency. As you saw in the previous section, a filter's response can be strongly influenced by the phase of the input signal. To determine response at a particular frequency, you can test all phases at that frequency. Alternatively, provided the filter is linear, you can present just two signals – a cosine wave at the test frequency and a sine wave at the same frequency. The filter's amplitude response is the magnitude of the vector sum of the two.

See *Linearity* on page 16.

Analytic and numerical procedures called *transforms* can be used to determine frequency response. The *Laplace transform* is appropriate for continuous functions, such as signals in the analog domain. The *Fourier transform* is appropriate for signals that are sampled periodically or signals that are themselves periodic. A variant intended for computation on data that has been sampled is the *discrete Fourier transform* (DFT). An elegant scheme for numerical computation of the DFT is the *fast Fourier transform* (FFT). The *z-transform* is essentially a Fourier transform whose frequency axis wraps around the unit circle, instead of being expressed as a cartesian coordinate. All of these transforms represent mathematical ways to determine a system's response to cosine waves and sine waves over a range of frequencies. The result of a transform is an expression or graph in terms of frequency.

Ronald N. Bracewell, *The Fourier Transform and its Applications*, Second Edition. New York: McGrawHill, 1985.

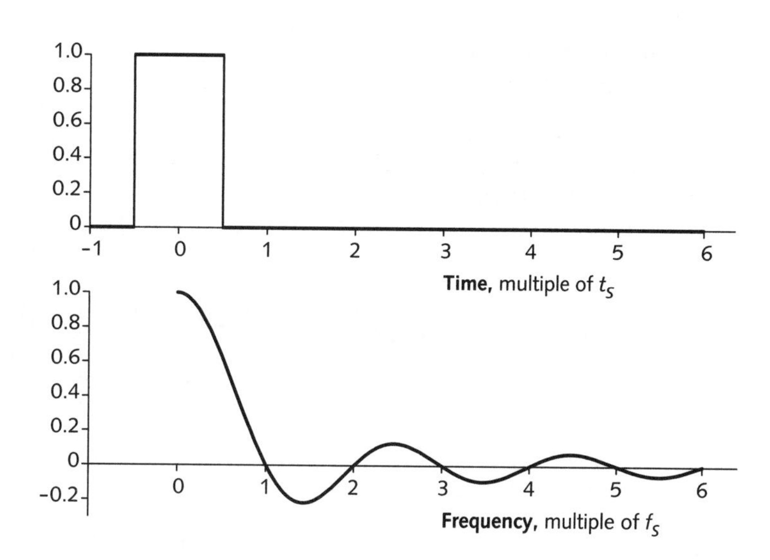

Figure 3.5 **Frequency response of a boxcar filter.** The top graph shows a boxcar waveform, symmetrical around $t = 0$. Its frequency spectrum, a *sinc* function, is shown underneath, plotted in single-sided fashion for positive frequencies only. Conceptually and mathematically, the response can be considered to extend symmetrically to negative frequencies.

Frequency response of a boxcar

$$\text{sinc } x = \frac{\sin(\pi x)}{\pi x}$$

The function *sinc* is unrelated to *sync*, synchronization.

Figure 3.5 above shows, in the top graph, the weighting function of *Boxcar filtering* on page 46. The Fourier transform of the boxcar function – that is, the frequency response of a boxcar weighting function – takes the shape of (sin x)/x, pronounced *sine x over x*. This function is so important that – with its argument scaled by pi – it has been given the special symbol *sinc*, pronounced *sink*. It is graphed in Figure 3.5 above.

A presampling filter needs to severely attenuate frequencies at and above half the sample rate. You can see that this requirement is not met by a boxcar weighting function. The graph of sinc predicts frequencies where aliasing can be introduced. Figure 3.4 showed an example of a sinewave at $0.75f_S$; the value of sinc at 0.75 shows that aliasing will result.

You can gain an intuitive understanding of the boxcar weighting function by considering that when the input frequency is such that an integer number of cycles lie under the boxcar, the response will be null. But when an integer number of cycles, plus a half-cycle, lie under the weighting function, the response will exhibit a local maximum that can admit an alias.

Figure 3.6 **The sinc function** is the ideal temporal weighting function – or *impulse response* – for a lowpass filter, but its infinite extent makes it physically unrealizable. Practical digital lowpass filters employ approximations of sinc.

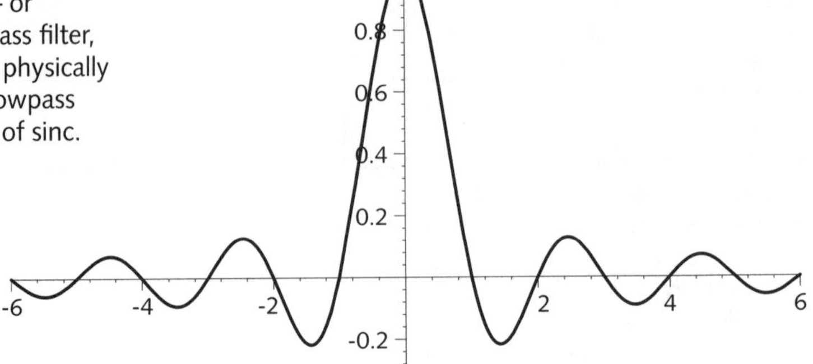

The temporal weighting functions used in video are usually symmetrical; nonetheless, they are usually graphed in a two-sided fashion. The frequency response of a filter suitable for real signals is symmetrical about zero; conventionally, frequency response is graphed in one-sided fashion starting at zero frequency ("DC"). Sometimes it is useful to consider or graph frequency response in two-sided style. Figure 3.6 above shows a two-tailed graph of the sinc function.

The response of an analog filter is a function of frequency on the positive real axis; in theory there is no upper bound on frequency. But in a digital filter the response to a test frequency f_T is identical to the response at f_T offset by any integer multiple of the sampling frequency: The frequency axis "wraps" at multiples of the sampling rate. Sampling theory also dictates "folding" around half the sample rate. Frequency response in the digital domain is usually graphed from zero to half the sampling rate.

Frequency response of point sampling

Taking an instantaneous sample of a waveform is equivalent to using a weighting function that is unity at the sample instant, and zero everywhere else. Casting point sampling in this framework allows use of the Fourier transform. The Fourier transform of an impulse function is constant, unity, at all frequencies.

A signal sampled by this function, without filtering, will admit aliases equally from all input frequencies.

To obtain a presampling filter that rejects potential aliases, we need to pass low frequencies, up to almost half the sample rate, and reject frequencies above it. We need a frequency response that is constant at unity up to just below $0.5f_S$, whereupon it drops to zero. We need a filter function whose *frequency* response – not time response – resembles a boxcar.

Remarkably, the Fourier transform possesses the mathematical property of being its own inverse. The Fourier transform of a boxcar *frequency* response produces a *sinc* function in the time domain: *sinc* is the ideal temporal weighting function for use in a presampling filter. But there are several theoretical and practical difficulties in using sinc. In practice, we use approximations of that function.

Fourier transform pairs

Figure 3.7 overleaf shows Fourier transform pairs for several different functions. In the left column is a set of waveforms; beside each waveform is its frequency spectrum. Functions that have short time duration transform to functions with widely distributed frequency components. Conversely, functions that are compact in their frequency representation transform to temporal functions with long duration.

Gaussian function:

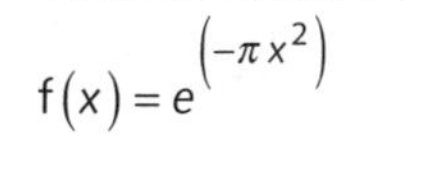

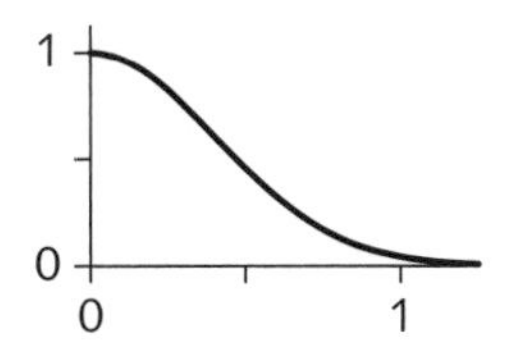

A Gaussian function – the middle transform pair in Figure 3.7 – has the unique property of transforming to itself through the Fourier transform. A Gaussian function has moderate spread in both the time domain and in the frequency domain; it has infinite extent, but becomes negligibly small more than a few units from the origin. A Gaussian can be considered to lie at the balance point between the distribution of power in the time domain and the distribution of power in the frequency domain.

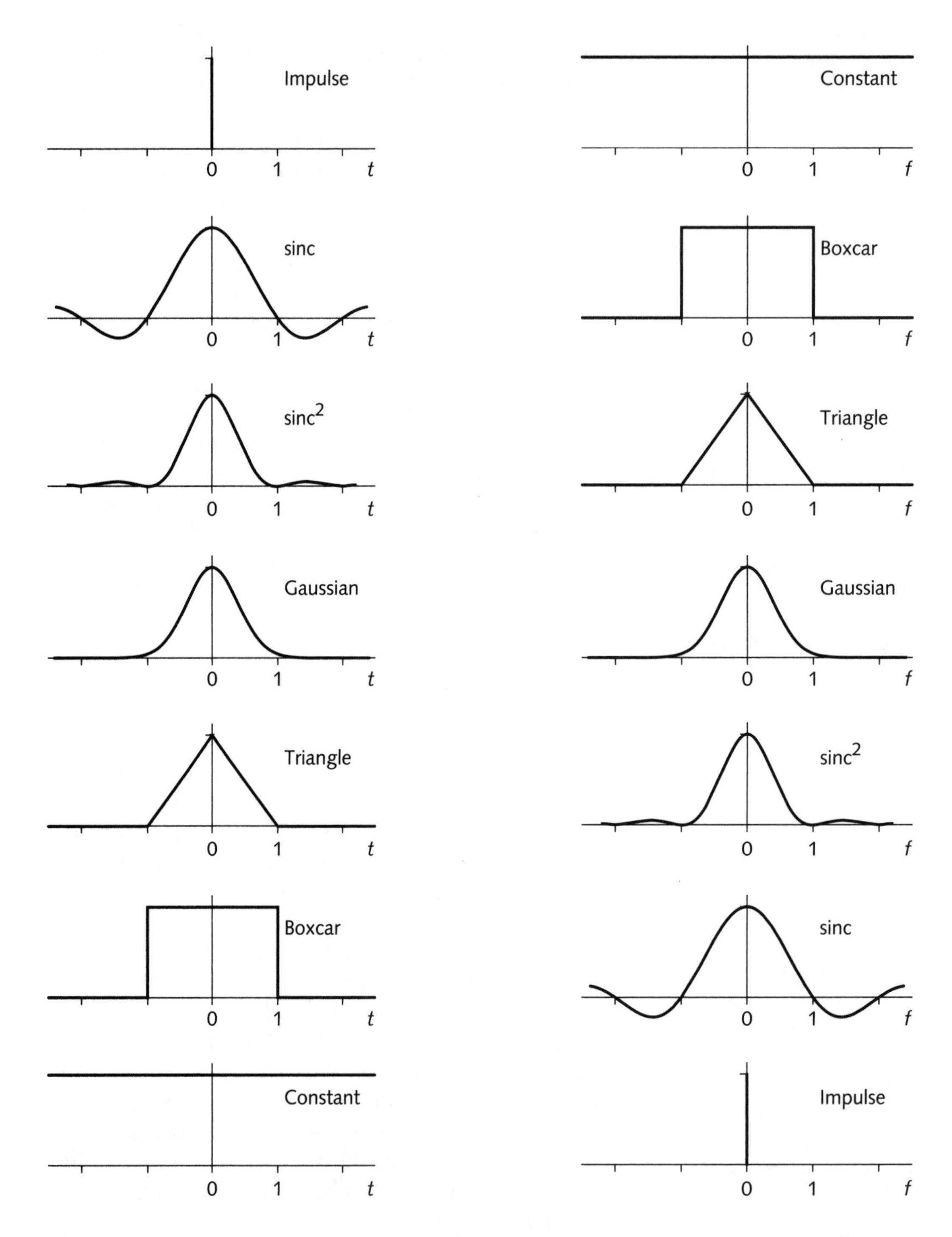

Figure 3.7 **Fourier transform pairs** for several different functions are shown in these graphs. In the left column is a set of waveforms in the time domain; beside each waveform is its frequency spectrum.

Figure 3.8 **Response of (1, 1) FIR filter.** This graph shows the frequency response of a digital filter that computes the sum of two adjacent samples.

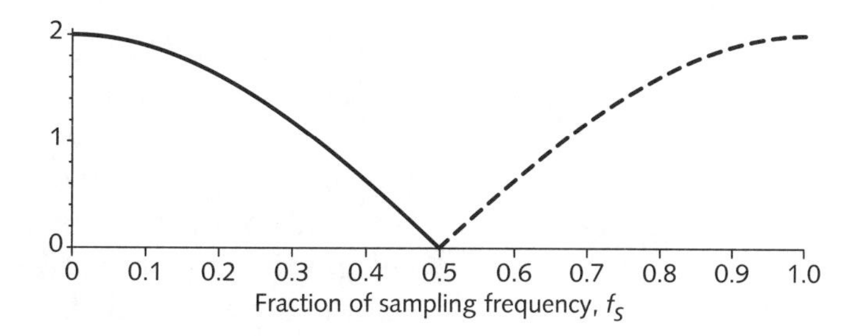

Digital filters

Filtering is necessary prior to digitization, to bring a signal into the digital domain without aliases. I have described filtering as integration using different weighting functions; an antialiasing filter performs the integration using analog circuitry. But once digitized, a signal can be filtered directly in the digital domain by computing a weighted sum of samples.

Perhaps the simplest digital filter is one that just sums adjacent samples; the weights in this case are (1, 1). The frequency response of a (1, 1) filter is shown in Figure 3.8 above. It offers minimal attenuation to very low frequencies; as signal frequency approaches half the sampling rate, the response follows a cosine curve to zero. The gain of this filter is 2, the sum of its coefficients. Usually, a filter's output, or its coefficients, are scaled to produce an overall gain of unity.

Figure 3.9 below shows the response, scaled to unity, of a filter that adds a sample to the second previous sample, disregarding the central sample. The weights in this case are (1, 0, 1). This forms a *notch filter*, with a null at one quarter the sampling frequency.

Figure 3.9 **Response of (1, 0, 1) FIR filter.** This is the frequency response of a filter that takes the average of a sample and the second preceding sample, ignoring the sample in between.

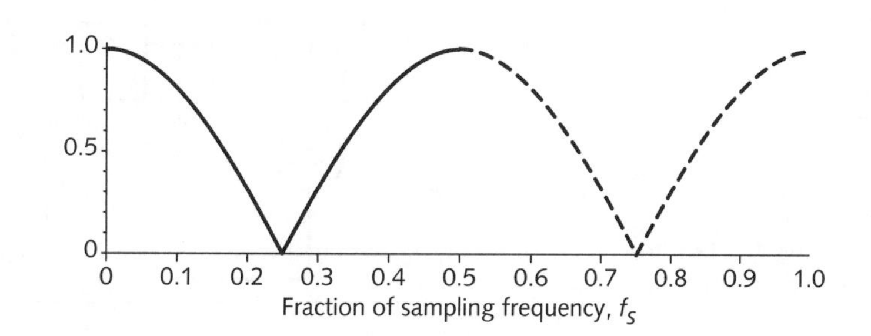

Impulse response

I have explained filtering as weighted integration along the time axis. I coined the term *temporal weighting function* to denote the weights. I consider my explanation of filtering in terms of its operation in the temporal domain to be more intuitive to a digital technologist than a more conventional explanation that starts in the frequency domain. But my term *temporal weighting function* is nonstandard, and I must now introduce the widely accepted but unintuitive term *impulse response*.

A digital impulse signal is a single sample having unity amplitude amidst a stream of zeros. An analog impulse signal has infinitesimal duration, infinite amplitude, and an integral of unity. The *impulse response* of a filter is its response to an input that is identically zero except for an instant at $t=0$, when it is unity.

Finite impulse response (FIR) filters

A digital filter generally implements temporal weighting directly. On page 53, I described two very simple digital filters, where every coefficient was either zero or one. Figure 3.10 below shows the block diagram of a filter having fractional coefficients; the weighting is implemented using multipliers. The blocks labeled *R* form a shift register. When a digital impulse is presented to this filter, the weighting coefficients are scanned out in turn: The set of coefficients and the impulse response are identical. The impulse response is finite in extent, because a finite number of positions encompass the nonzero coefficients.

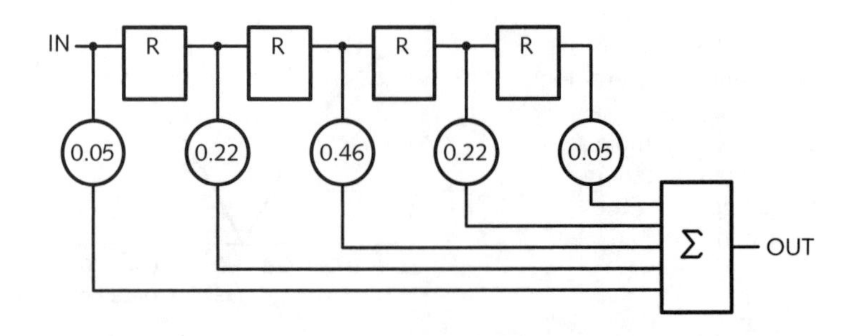

Figure 3.10 **FIR filter block diagram.**

Symmetry:
$f(x) = f(-x)$

Antisymmetry:
$f(x) = -f(-x)$

I have described impulse responses that are symmetrical around an instant in time. You might think $t=0$ should denote the beginning of time, but it is convenient to shift the time axis so that $t=0$ corresponds to the central point of a filter's impulse response. A *finite impulse response* (FIR) filter has a limited number of coefficients that are nonzero. When the input impulse lies outside this interval, the response is zero. With a few important exceptions, most digital filters used in video are FIR filters, and most have impulse responses either symmetric or antisymmetric around $t=0$.

Physical realizability of a filter

In order to be implemented, a digital filter must be *physically realizable:* It is a practical necessity to have a temporal weighting function of limited duration. An FIR filter requires storage of several input samples, and it requires several multiplication operations to be performed during each sample period. The number of input samples stored is called the *order* of the filter, or its number of *taps*. A straightforward technique can be used to exploit the symmetry of the impulse response to eliminate half the multiplications. If a particular filter has fixed coefficients, the multiplications can be performed by table lookup.

Here I use the word *truncation* to indicate the forcing to zero of a filter's weighting function beyond a certain point. The nonzero coefficients in a weighting function may involve theoretical values that have been quantized to a certain number of bits. This *coefficient quantization* can be accomplished by *rounding* or by *truncation*. Be careful to distinguish the two kinds of truncation.

When a temporal weighting function is truncated past a certain point, its transform – its frequency domain representation – will have unlimited extent. The craft of filter design involves carefully choosing the order of the filter – that is, the position beyond which the weighting function is made zero. That position needs to be far enough from the center that the filter's high-frequency response is small enough to be negligible for the application.

Signal processing accommodates the use of impulse responses having negative values, and negative coefficients are common in digital signal processing. But image capture and image display involve sensing and generating light, which cannot have negative power, so negative weights cannot always be realized. If you study the transform pairs on page 52 you will see that your ability to tailor the frequency response of a filter is severely limited when you cannot use negative weights.

I have been describing Fourier transforms of impulse responses that are appropriate for presampling filters. It is possible to implement a boxcar filter directly in the analog domain. But the implementation of an analog filter generally bears little relationship to its impulse response. Analog filters rarely implement temporal weighting directly. An analog filter more complex than a boxcar is best described in terms of a Laplace transform, not a Fourier transform; and an analog filter is usually specified in terms of its frequency response, not in terms of its impulse response. Despite the major conceptual and implementation differences, analog filters and FIR filters – and *IIR* filters, to be described – are all characterized by their frequency response.

An analog filter performs integration by storing a magnetic field in an inductor (coil) using the electrical property called *inductance* (L), or by storing an electrical charge in a capacitor using the electrical property called *capacitance* (C). In low-performance filters, *resistance* (R) is used as well. An ordinary analog filter has an impulse responses that is infinite in temporal extent.

Phase response (group delay)

Until now I have described the *frequency* response of filters. *Phase* response is also important. Consider a symmetrical FIR filter having 15 taps. No matter what the input signal, the output will have an effective delay of 8 sample periods, corresponding to the central sample of the filter's impulse response. The time delay of an FIR filter is constant, independent of frequency.

125 ns, 45° at 1 MHz

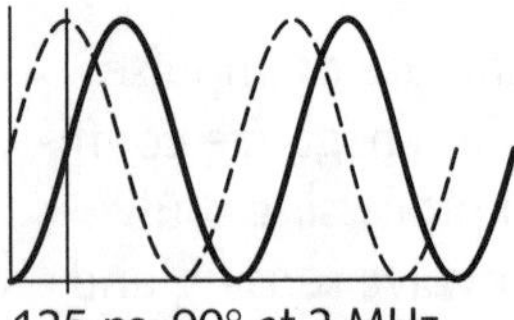

125 ns, 90° at 2 MHz

Figure 3.11 **Linear phase.**

Consider a sine wave at 1 MHz, and a second sine wave at 1 MHz but delayed 125 ns. This delay could be expressed as a phase shift of 45°. But if the time delay remains constant and the frequency doubles, the phase offset doubles to 90°. With constant time delay, phase offset increases in direct (linear) proportion to the increase in frequency. For obscure reasons this condition is called *constant group delay*. Since in this condition phase delay is directly proportional to frequency, its synonym is *linear phase*.

It is a characteristic of many filters that delay varies somewhat as a function of frequency. Symmetric FIR filters exhibit linear phase. Since linear phase is a highly desirable property in a video system, FIR filters are preferred over other sorts of filters.

Infinite Impulse Response (IIR) filters

The digital filters described so far have been members of the FIR class. A second class of digital filter is characterized by having a potentially *infinite impulse response* (IIR). An IIR or *recursive* filter computes an output sample by computing a weighted sum of input samples – as is the case in an FIR filter – but adds to this a weighted sum of previous *output* samples. An IIR filter cannot possess exactly linear phase, although a complex IIR filter can be designed to have arbitrarily low phase error. Because IIR filters usually have poor phase response, they are not ordinarily used in video.

Compensation of undesired phase response in a filter is known as *equalization*. This is unrelated to the *equalization* pulses that form part of sync.

Lowpass filter

A lowpass filter lets low frequencies pass undisturbed, but attenuates high frequencies. Figure 3.12 overleaf characterizes a lowpass filter. The response has a *passband*, where the filter's response is nearly unity; a *transition band*, where the response has intermediate values; and a *stopband*, where the filter's response is nearly zero. The *cutoff frequency* – or *bandwidth*, f_C – is defined as the frequency where the response of the filter has fallen to half its magnitude at a reference frequency (usually zero).

The passband is characterized by the passband edge frequency f_P and the passband ripple ∂_P (sometimes denoted ∂_1). The stopband is characterized by its edge frequency f_S and ripple ∂_S (sometimes denoted ∂_2). The transition band between f_P and f_S has width Δf.

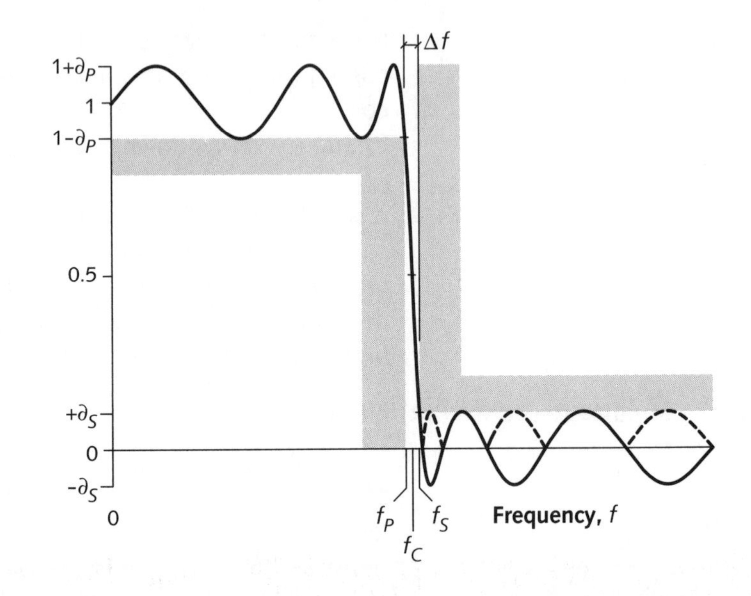

Figure 3.12 **Lowpass filter characterization.** A lowpass filter for use in video sampling or reconstruction has a cutoff frequency f_C, where the attenuation is 0.5. In the *passband*, response is unity within ∂_P, usually 1 percent or so. In the *stopband*, response is zero within ∂_S, usually 1 percent or so. The *transition band* lies between the edges of the passband and stopband. The solid line shows that at certain frequencies, the filter causes phase inversion. Filter responses are usually plotted as magnitudes, so phase inversion in the stopband is reflected as the absolute values shown in dashed lines.

The complexity of a lowpass filter is roughly determined by its *transition ratio* $\Delta f/f_C$ (or in the case of a digital filter, $\Delta f/f_S$). The narrower the transition band, the more complex the filter. Also, the smaller the ripple in either the passband or the stopband, the more complex the filter.

In analog filter design, frequency response is generally graphed in log-log coordinates, with frequency in hertz (Hz) and amplitude in decibels (dB). In digital filter design, frequency is usually graphed linearly from zero to half the sampling frequency. The passband and stopband response of a digital filter are usually graphed logarithmically; the passband response is usually magnified to emphasize small departures from unity.

The templates standardized in Rec. 601 for studio digital video are shown in Figure 3.13 opposite.

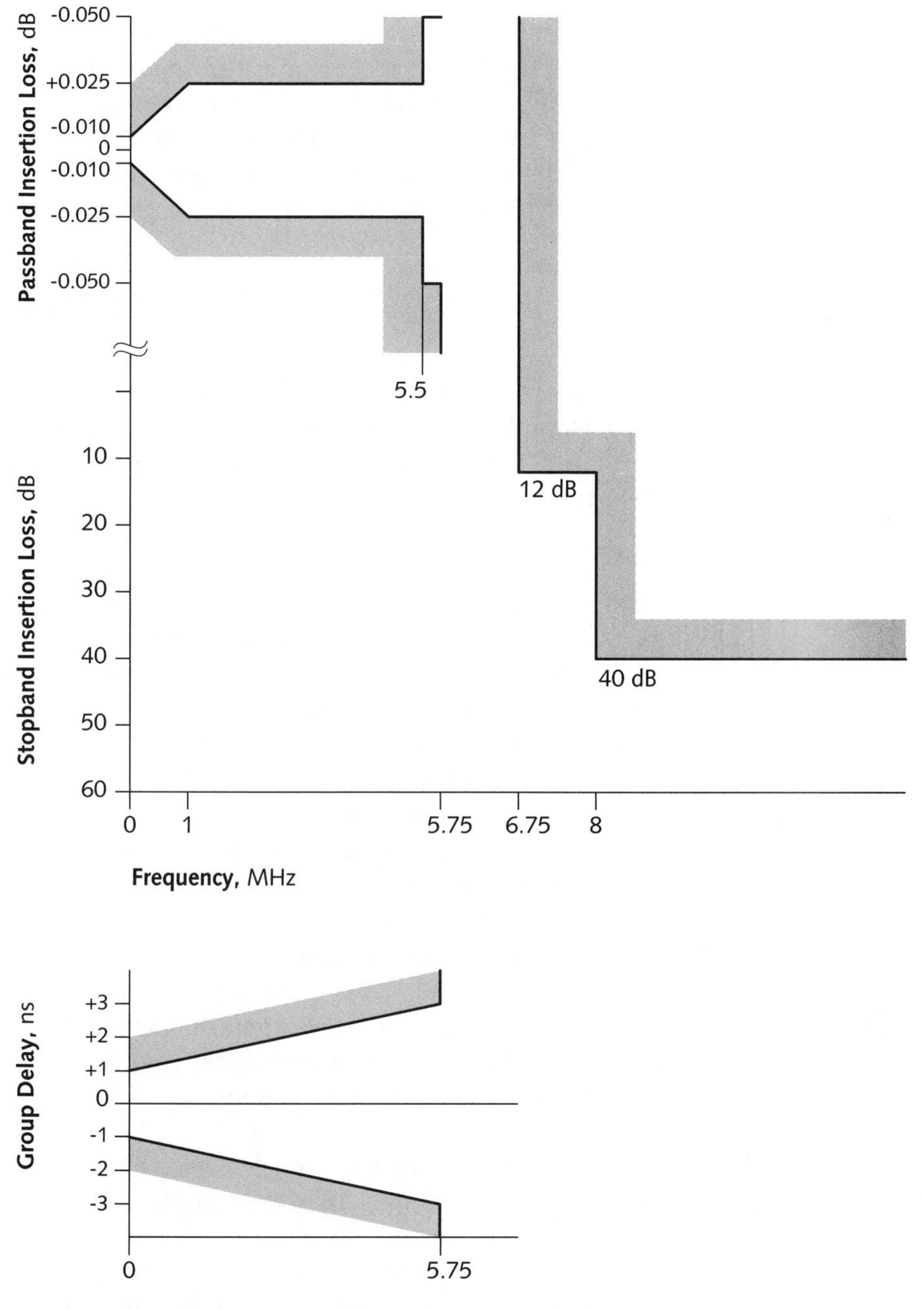

Figure 3.13 **Rec. 601 filter templates.** These filter templates are standardized for studio digital video systems in Rec. 601-4. The top template shows frequency response, detailing the passband (at the top) and the stopband (in the middle). The bottom template shows the group delay specification.

Digital filter design

The easiest way to design a digital filter is to use weighting coefficients that comprise an appropriate number of point-samples of a theoretical impulse response. Coefficients beyond a certain point – the *order* of the filter – are simply omitted. Equation 3.1 implements a 9-tap filter that approximates a Gaussian:

Eq 3.1

$$y_i = \frac{1x_{i-4} + 9x_{i-3} + 43x_{i-2} + 110x_{i-1} + 150x_i + 110x_{i+1} + 43x_{i+2} + 9x_{i+3} + 1x_{i+4}}{476}$$

Omission of coefficients causes frequency response to depart from the ideal. If the omitted coefficients are much greater than zero, actual frequency response can depart significantly from the ideal.

Figure 3.14 **FIR filter example, 23-tap lowpass.**

$$\begin{aligned} y_i = {} & 0.098460\,x_{i-12} \\ & +0.009482\,x_{i-11} \\ & -0.013681\,x_{i-10} \\ & +0.020420\,x_{i-9} \\ & -0.029197\,x_{i-8} \\ & +0.039309\,x_{i-7} \\ & -0.050479\,x_{i-6} \\ & +0.061500\,x_{i-5} \\ & -0.071781\,x_{i-4} \\ & +0.080612\,x_{i-3} \\ & -0.087404\,x_{i-2} \\ & +0.091742\,x_{i-1} \\ & +0.906788\,x_{i} \\ & +0.091742\,x_{i+1} \\ & -0.087404\,x_{i+2} \\ & +0.080612\,x_{i+3} \\ & -0.071781\,x_{i+4} \\ & +0.061500\,x_{i+5} \\ & -0.050479\,x_{i+6} \\ & +0.039309\,x_{i+7} \\ & -0.029197\,x_{i+8} \\ & +0.020420\,x_{i+9} \\ & -0.013681\,x_{i+10} \\ & +0.009482\,x_{i+11} \\ & +0.098460\,x_{i+12} \end{aligned}$$

An improvement on this brute-force approach is *windowed filter design*, where instead of abrupt omission, coefficients from the ideal impulse response are "windowed" (weighted) by a function that peaks at unity at the center of the filter, and diminishes gently to zero at the extremities of the interval. Several different window functions are commonly used. Windowing offers smoother frequency response than design by truncation, but still offers relatively poor control of the response.

Few closed-form methods are known to design optimum digital filters. Design of a high-performance filter usually involves successive approximation, optimizing by trading design parameters back and forth between the time and frequency domains.

The implementation of a high-quality video filter for video is shown in Figure 3.14 in the margin.

The response of a practical lowpass filter of studio-quality performance is shown in Figure 3.15 opposite; its cutoff frequency is $0.25 f_S$. A consumer filter might have ripple two orders of magnitude worse.

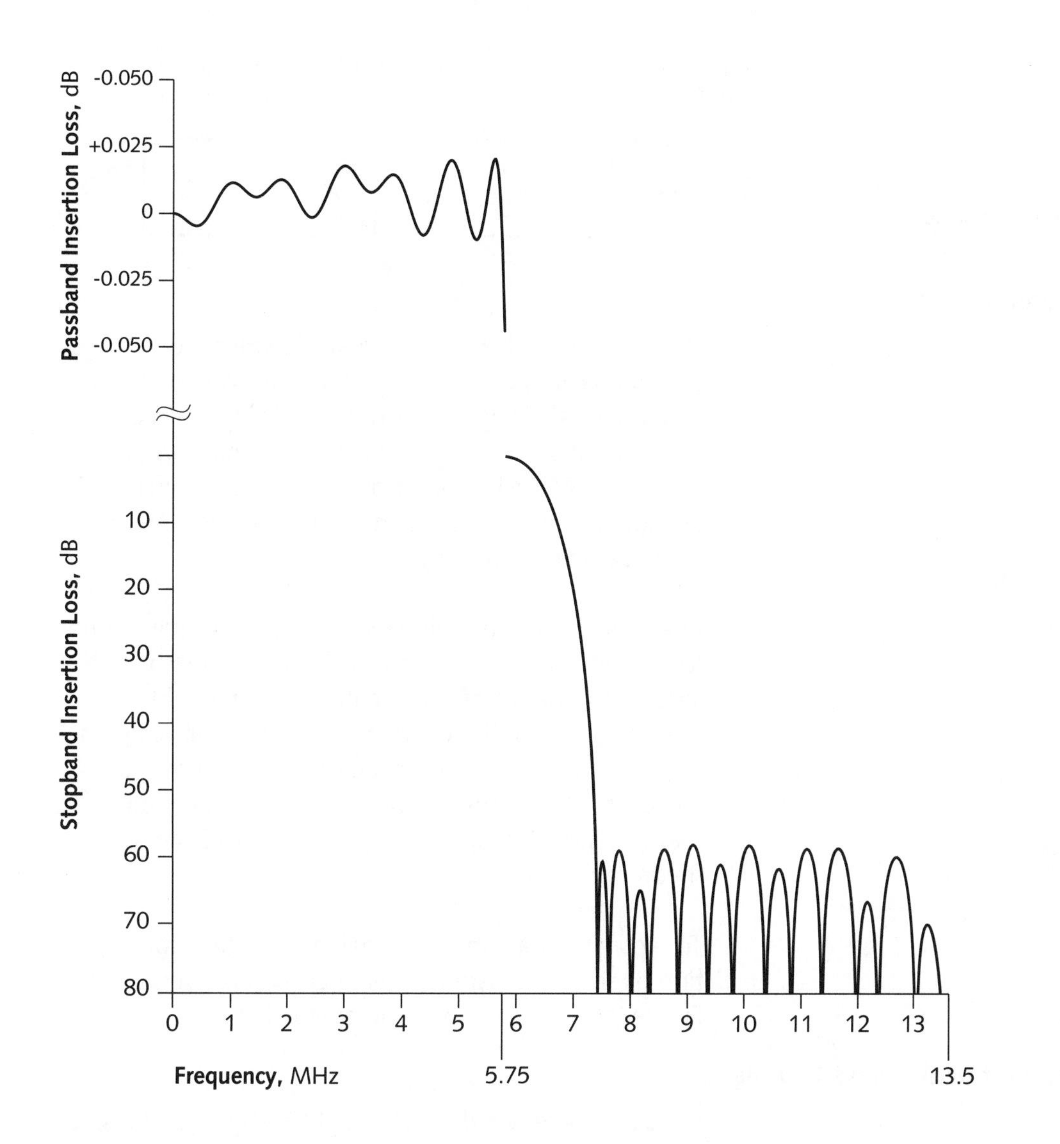

Figure 3.15 **Half-band filter.** This graph shows the frequency response of a practical filter whose cutoff is at one-quarter its sampling frequency of 27 MHz. The graph is linear in the ordinate (frequency) and logarithmic in the abscissa (response). The top portion shows that the passband has an overall gain of unity and a uniformity (*ripple*) of about ±0.02 dB: In the passband, its gain varies between about 0.997 and 1.003. The bottom portion shows that the stopband is rejected with an attenuation of about -60 dB: The filter has a gain of about 0.001 at these frequencies. This data, for the GF9102A halfband filter, was kindly provided by Gennum Corporation.

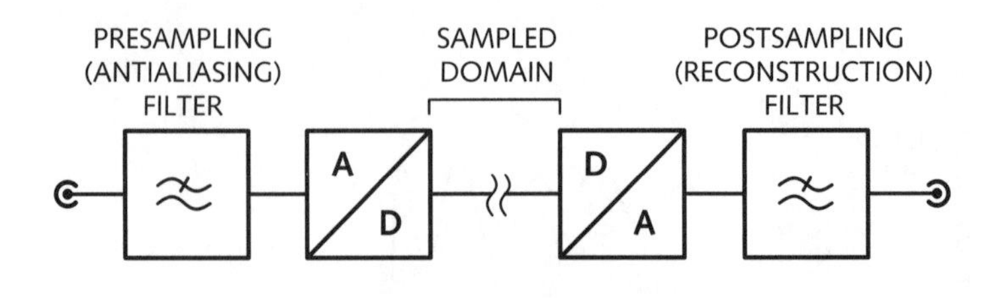

Figure 3.16 **Sampling and reconstruction.**

Reconstruction

Digitization involves sampling and quantization. These operations are performed in an analog-to-digital converter (ADC). Whether the signal is quantized then sampled, or sampled then quantized, is relevant only within the ADC: The order of operations is immaterial outside that subsystem. Contemporary video ADCs quantize first, then sample.

I have explained that filtering is generally required prior to sampling in order to avoid the introduction of aliases. Avoidance of aliasing in the sampled domain has obvious importance. In order to avoid aliasing, an analog presampling filter needs to operate prior to analog-to-digital conversion. If aliasing is avoided, then the sampled signal can, according to Shannon's theorem, be reconstructed.

To reconstruct an analog signal, an analog reconstruction filter is necessary following D-to-A conversion. Figure 3.16 above shows the overall flow.

Reconstruction close to $0.5f_S$

Consider the example in Figure 3.17 opposite of a sine wave at $0.44f_S$. This signal meets the sampling criterion, and can be perfectly represented in the digital domain. But from an intuitive point of view, it is difficult to predict the underlying sinewave from samples 3, 4, 5, and 6 in the lower graph. When reconstructed using a Gaussian, this high frequency signal vanishes. To be reconstructed accurately, a waveform with a significant amount of power near half the sampling rate must be reconstructed with a high quality filter. This waveform is accurately reconstructed by a high quality filter, such as that of Figure 3.14, on page 60.

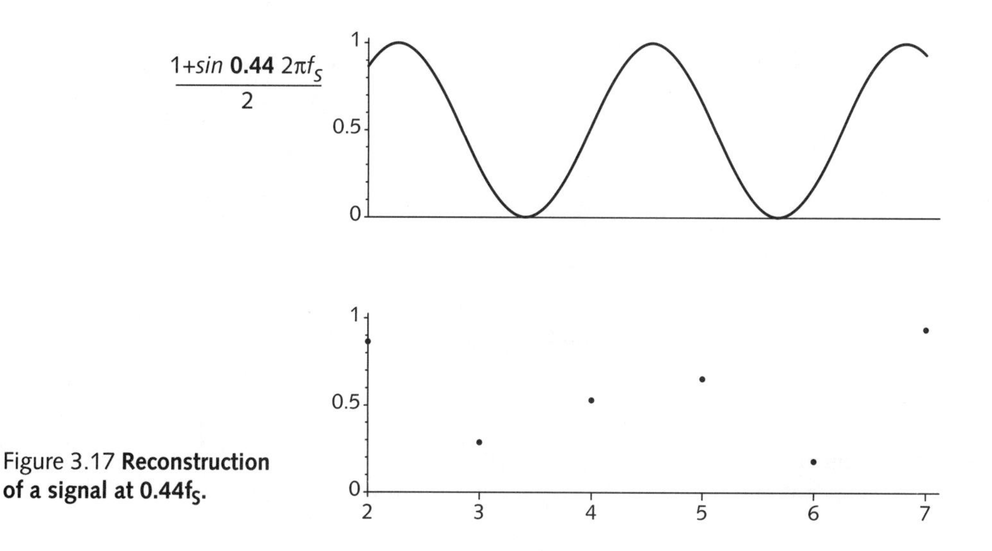

Figure 3.17 **Reconstruction of a signal at $0.44f_S$.**

(sin x)/x correction

I have described how it is necessary for an analog reconstruction filter to follow digital-to-analog conversion. A classic lowpass filter would suffice if the DAC produced an impulse waveform where the amplitude of each impulse was modulated by the corresponding code value: All would be well if the DAC output resembled my "point" graphs, with power at the sample instants and no power in between. Recall that a waveform comprising just unit impulses has uniform frequency response across the entire spectrum.

Unfortunately, a typical DAC does not produce an impulse waveform for each sample. Instead, each converted sample value is held for the entire duration of the sample: A typical DAC produces a boxcar waveform in time. A boxcar function has a frequency response that is described by sinc. It would be impractical to have a DAC with an impulse response, because an analog filter responds to roughly the integral of the signal, and the amplitude of the impulses would have to be impractically high for the integral of the impulses to match the integral of a boxcar function.

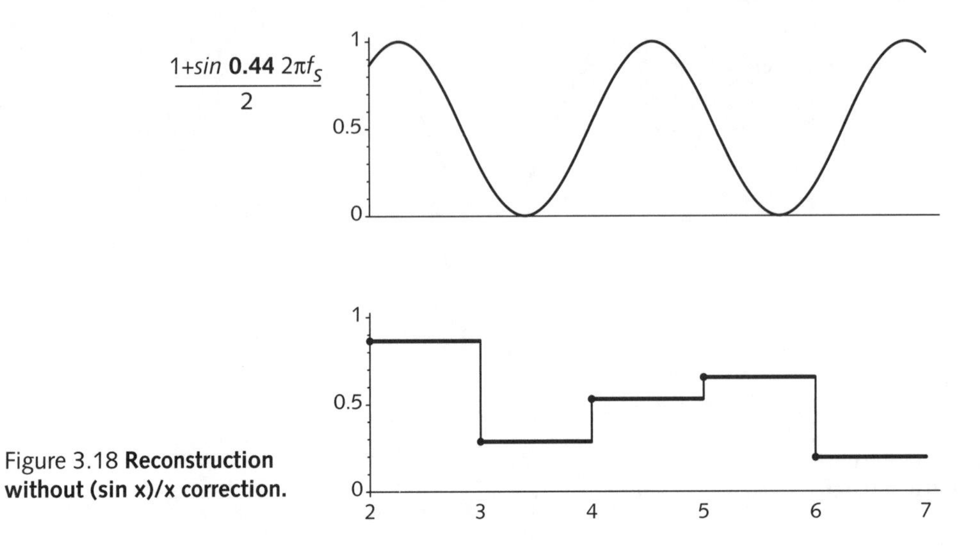

Figure 3.18 **Reconstruction without (sin x)/x correction.**

In Figure 3.18 above, the top graph is a sine wave at $0.44f_S$; the bottom graph shows the boxcar waveform produced by a conventional DAC. Even with a high quality reconstruction filter, whose response extends close to half the sampling rate, it is evident that reconstruction by a boxcar function reduces the amplitude of high frequency components of the signal.

The DAC's holding of each sample value throughout the duration of its sample interval corresponds to a filtering operation, with a frequency response of (sin x)/x. The top graph of Figure 3.19 opposite shows the attenuation due to this phenomenon.

This could be called *sinc correction*, but it isn't.

The effect is overcome by (sin x)/x *correction:* The frequency response of the reconstruction filter is modified to include peaking corresponding to the reciprocal of (sin x)/x. In the passband, the filter's response increases gradually to about 4 dB as a function of frequency to compensate the loss. Above the passband edge frequency, the response of the filter must decrease rapidly to produce a large attenuation near half the sampling frequency, to provide alias-free

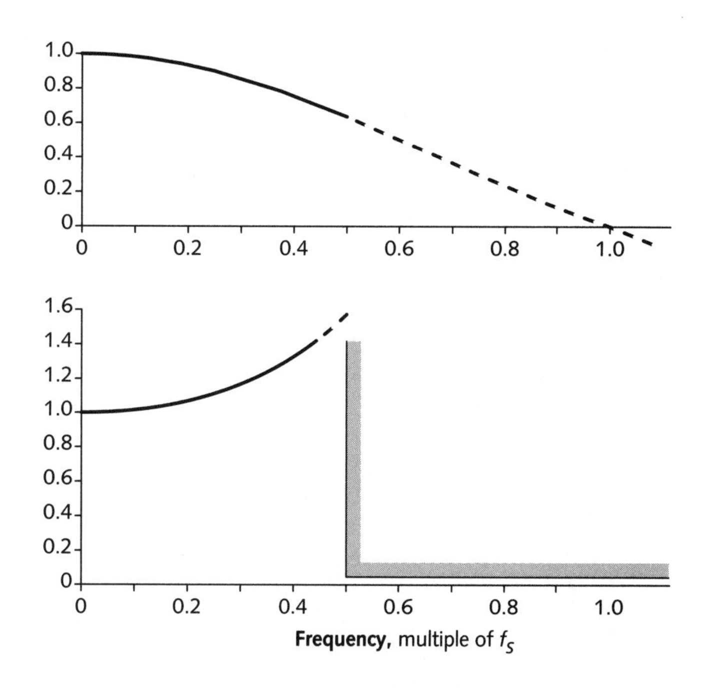

Figure 3.19 **(sin x)/x correction** is necessary following digital-to-analog conversion. A conventional DAC has a boxcar output waveform having a frequency response shown in the upper graph. The lower graph shows the response of the (sin x)/x correction filter necessary to compensate this falloff.

reconstruction. The bottom graph of Figure 3.19 shows the idealized response of a filter having (sin *x*)/*x* correction.

In this chapter, I have detailed one-dimensional filtering. The following chapter, *Image digitization and reconstruction,* introduces two- and three-dimensional sampling and filters.

Further reading

C. Britton Rorabaugh, *Digital Filter Designer's Handbook.* Blue Ridge Summit, PA: TAB Books div. McGraw-Hill, 1993.

Sanjit K. Mitra and James F. Kaiser, *Handbook for Digital Signal Processing.* New York: John Wiley & Sons, 1993.

For an approachable introduction to digital signal processing (DSP) from a programmer's point of view, see Rorabaugh. His book includes the source code for programs to design filters – that is, to evaluate filter coefficients. For comprehensive and theoretical coverage of DSP, see Mitra and Kaiser.

Image digitization and reconstruction 4

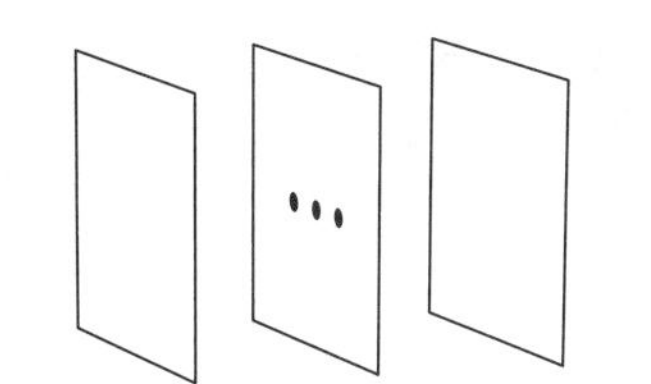
Figure 4.1 **Horizontal domain.**

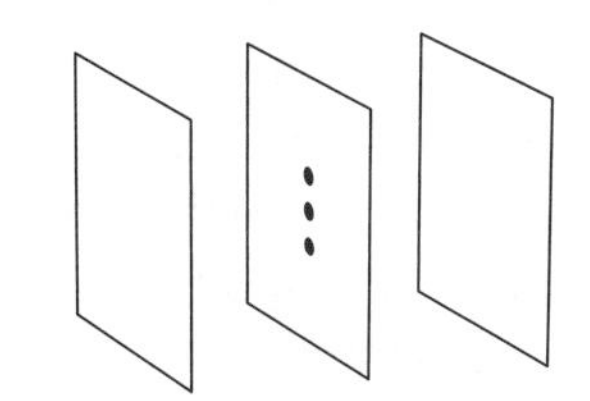
Figure 4.2 **Vertical domain.**

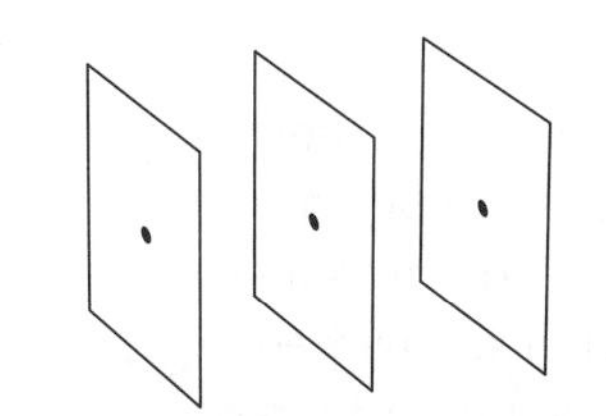
Figure 4.3 **Temporal domain.**

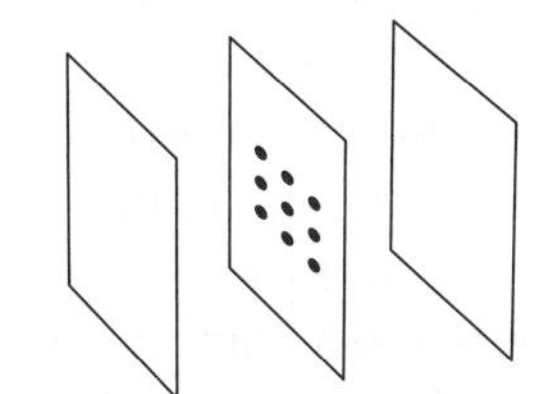
Figure 4.4 **Spatial domain.**

In Chapter 3, *Filtering and sampling*, on page 43, I described how to analyze a signal that is a function of the single dimension of time, such as an audio signal. Sampling theory also applies to a signal that is a function of one dimension of space, such as a single scan line of a video signal. This is the horizontal or *transverse* domain, sketched in Figure 4.1. If an image is scanned line by line, the waveform of each line can be treated as an independent signal. The techniques of filtering and sampling in one dimension, discussed in the previous chapter, apply directly to this case.

Now consider a set of points arranged vertically that originate at the same displacement along each of several successive scan lines, sketched in Figure 4.2. Those points can be considered to be sampled by the scanning process itself. Sampling theory can be used to understand the properties of these samples.

A third dimension is introduced when a succession of images is sampled temporally to represent motion, as sketched in Figure 4.3.

Complex filters can act on two axes simultaneously. Figure 4.4 illustrates spatial sampling. The properties of the entire set of samples are considered all at once, and cannot necessarily be separated into independent horizontal and vertical aspects.

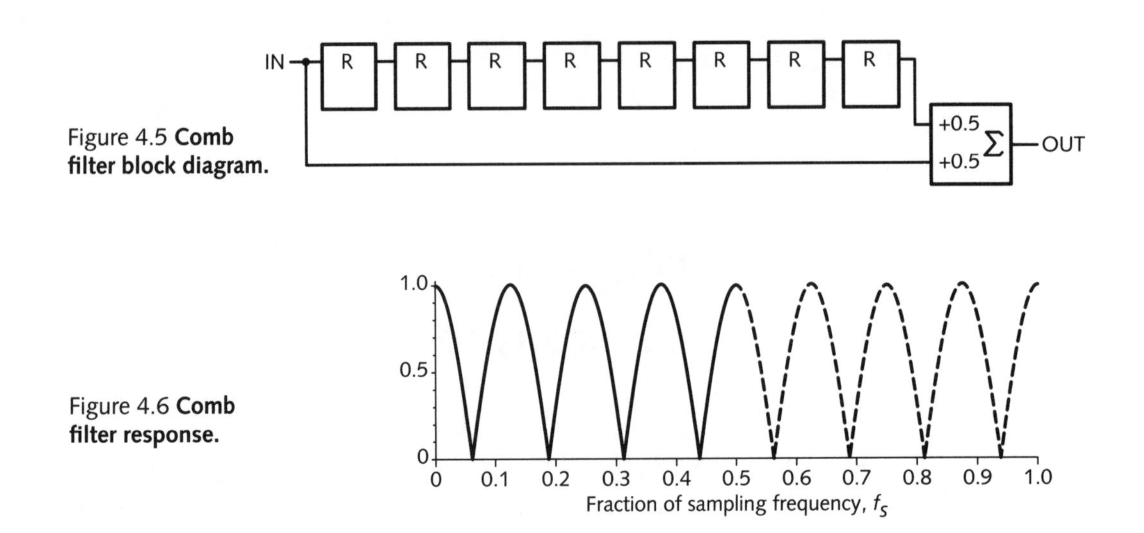

Figure 4.5 **Comb filter block diagram.**

Figure 4.6 **Comb filter response.**

Comb filtering

Figure 4.5 above shows the block diagram of an FIR filter having eight taps weighted 1, 0, 0, ..., 0, 1. The frequency response of this filter is shown in Figure 4.6 above; the response peaks when an exact integer number of cycles lie underneath the filter, and nulls when an integer-and-a-half cycles lie underneath. The peaks all have the same magnitude: The response is the same when exactly 1, 2, ..., or *n* samples are within its window. This graph is reminiscent of a comb, and the filter is called a *comb filter.*

You can consider this filter to operate in the single dimension of time. If the samples are taken from a sampled scan line from an image, the frequency response can represent horizontal spatial frequency, measured in cycles per picture width, instead of temporal frequency measured in cycles per second.

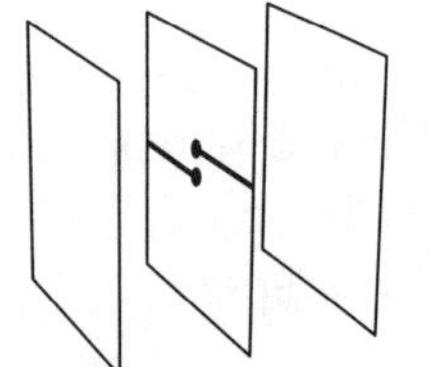

Figure 4.7 **Two samples, arranged vertically.**

Consider a sample from an image, and the sample immediately below, as sketched in Figure 4.7 in the margin. If the image has 858 samples per total line, and these two samples are presented to a comb filter having 857 zero-samples between the two "ones," then the action of the comb filter will be identical to the action of a filter having two taps weighted (1, 1)

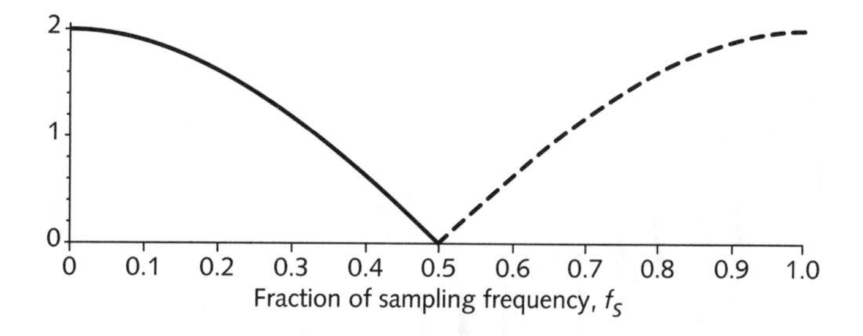

Figure 4.8 **Response of (1, 1) FIR filter.**

operating in the vertical direction. On page 53, I discussed the frequency response of a (1, 1) filter; Figure 4.8 above graphs its response.

Frequency spectrum of NTSC

In *Filtering and sampling*, on page 43, I explained how a one-dimensional waveform in time transforms to a one-dimensional frequency spectrum.

Consider an image where every scan line is identical. The time domain waveform of this signal is periodic at the line rate. Considering this signal in the one-dimensional frequency domain, all of the power of the signal is contained at multiples of the line rate: 0 (DC, or zero frequency), f_H (the line rate), $2f_H$, $3f_H$, and so on. The power in typical images tends to concentrate at low frequencies, and diminish towards higher frequencies.

If the content of successive scan lines varies slightly, the effect in the one-dimensional frequency spectrum is to broaden the spectral lines: power is centered at 0, f_H, $2f_H$, $3f_H$, and so on, but spreads somewhat into nearby frequencies. The luma component of a typical image is sketched at the top of Figure 4.9 overleaf.

You will see in *Composite NTSC and PAL*, on page 185, that color information in NTSC is encoded onto a subcarrier whose frequency is coherent with the line rate, having 227.5 cycles per total line. The NTSC subcarrier frequency is an odd multiple of half the line rate: its phase inverts at line rate, so subcarrier power is concentrated at $227.5f_H$.

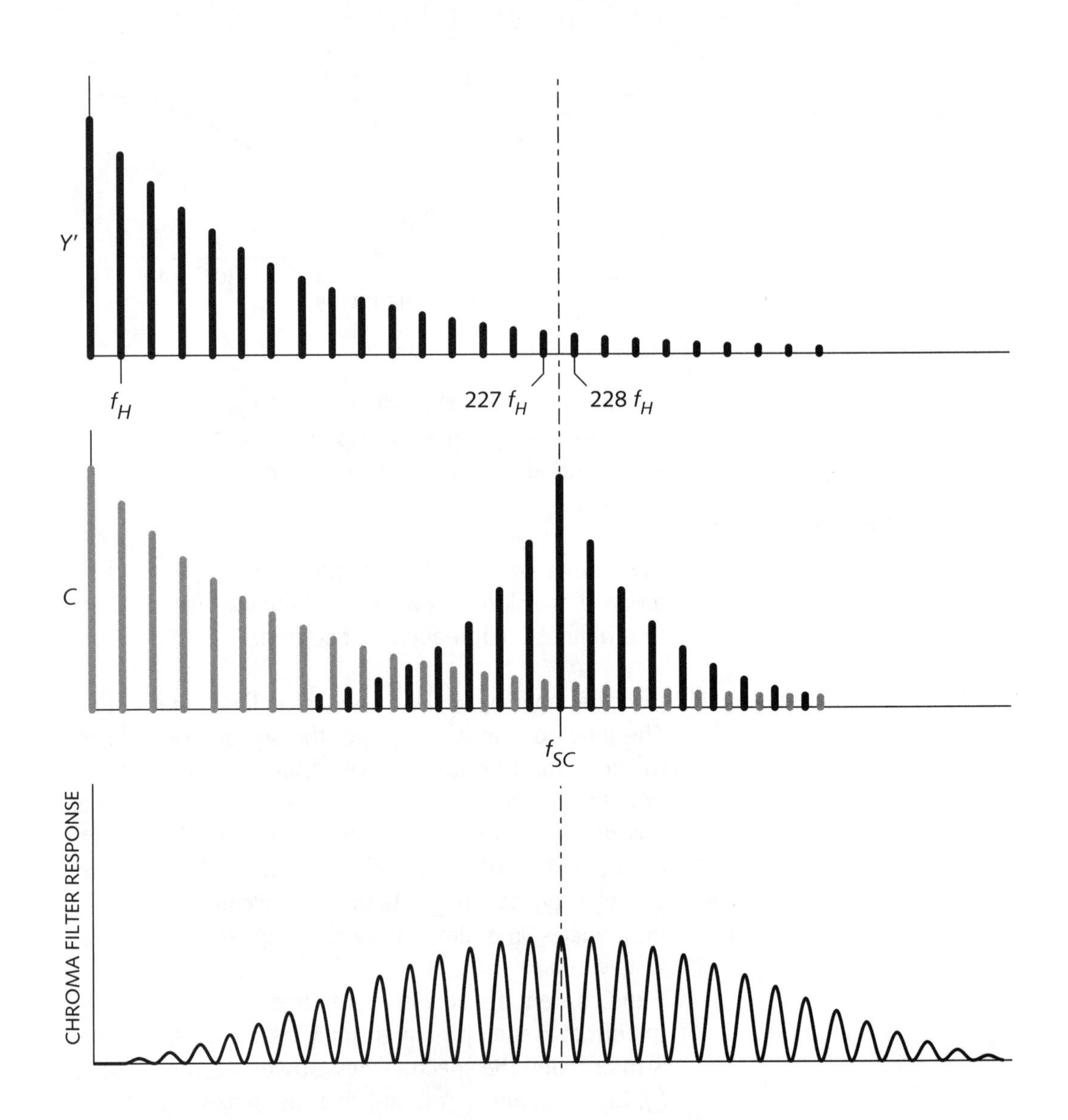

Figure 4.9 **Frequency spectrum of NTSC.** The top graph shows the spectrum of luma from a typical picture. Luma power clusters at integer multiples of the line rate (here exaggerated). The middle graph shows the spectrum of chroma after modulation onto the subcarrier, whose frequency is $227.5 f_H$, an odd multiple of half the line rate. Modulated chroma also clusters at multiples of the line rate, but the choice of subcarrier frequency causes modulated color information to cluster between the luma spectral lines. The bottom graph shows the frequency response of a comb filter in tandem with a bandpass filter; this combination separates chroma with minimal contamination from luma.

Consider a video signal where every scan line contains a sinewave of frequency f_T, but where the phase of the sinewave is inverted in alternate lines. This signal is periodic at half the line rate: All of the power in this signal lies at multiples of $f_H/2$. Furthermore, due to symmetry, the power lies only at odd multiples.

In composite NTSC and PAL video, color information is subject to a process called *quadrature modulation* that imposes color information onto a subcarrier; modulated chroma is then added to luma. I will detail this process in *Composite NTSC and PAL*, on page 185. Low frequency chroma information lies very near the subcarrier frequency; higher frequency color detail causes the modulated information to spread out in frequency around odd $f_H/2$ multiples, in the same manner that luma spreads from f_H multiples. The middle graph of Figure 4.9 shows the spectrum of modulated NTSC chroma in a typical image.

An NTSC decoder has to separate luma and chroma. One way to accomplish this is to use a *notch filter* operating in the horizontal domain that rejects signals in the region of subcarrier frequency. Figure 3.9, at the bottom of page 53, shows the response of a (1, 0, 1) filter, which has a notch at one quarter of the sampling frequency. This filter would reject chroma in a system sampled at four times the color subcarrier frequency ($4f_{SC}$). Although a notch filter eliminates chroma, a significant amount of luma that occupies similar horizontal frequencies is eliminated as well: A notch filter eliminates picture detail.

In *Comb filtering*, on page 68, I explained how to construct a filter that produces notches at regular intervals of the one-dimensional frequency spectrum. If a comb filter such as the one in Figure 4.5 has as many registers as there are samples per total line, the peaks will pass luma and the notches will reject modulated chroma. A comb filter rejects chroma as effectively as a notch filter, but it has the great advantage that luma detail in the range of subcarrier frequencies is retained.

If luma that has been separated from a composite signal is subtracted from the composite signal, chroma remains! Prior to subtraction, the composite signal must be delayed to compensate the delay of luma through the separation filter. The bottom graph of Figure 4.9 shows the chroma response of a comb filter cascaded with a bandpass filter.

Spatial frequency domain

In *Filtering and sampling*, on page 43, I explained how a one-dimensional waveform in time transforms to a one-dimensional frequency spectrum. This concept can be extended into two dimensions: The two dimensions of space can be transformed into two-dimensional spatial frequency. The content of an image can be expressed as horizontal and vertical spatial frequency components. Spatial frequency is graphed using *cycles per picture width* (C/PW) as an *x*-coordinate, and *cycles per picture height* (C/PH) as a *y*-coordinate. You can gain insight into the operation of an imaging system by exploring its spatial frequency response. In the following discussion, I will recast the explanation of the NTSC spectrum into the spatial frequency domain.

Figure 4.10 below shows an image comprising a sine-wave signal in the vertical direction. Four cycles occupy the height of the picture. The spatial frequency graph, to the right, shows that all of the power of the image is contained at coordinates (0, 4) of spatial frequency. In an image where each scan line takes a constant value, all of the power is located on the *y*-axis of spatial frequency.

Figure 4.10 **Spatial frequency domain.**

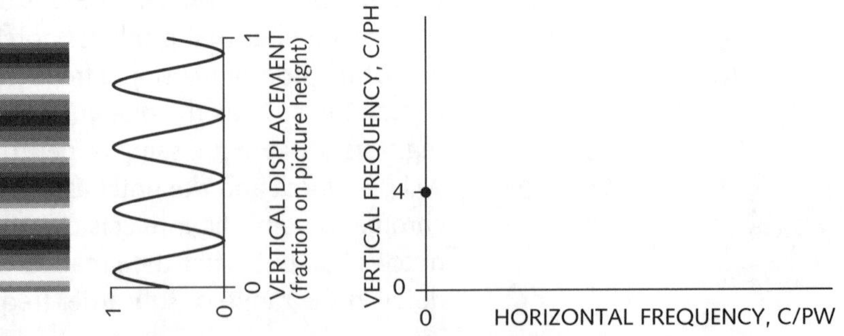

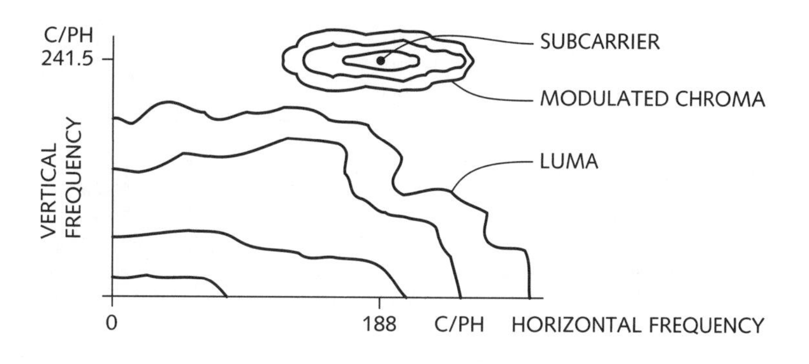

Figure 4.11 **Spatial frequency domain.**

If an image comprises scan lines with identical content, all of the power will be concentrated on the horizontal axis of spatial frequency. If the content of successive scans lines varies slightly, the power will spread to nonzero vertical frequencies. An image of diagonal bars would occupy a single point in spatial frequency, displaced from the *x*-axis and displaced from the *y*-axis.

Most video pictures have great similarity of picture information between successive rows and columns of sample, so the luma component of a video signal tends to cluster around spatial frequency coordinates (0, 0), at the lower left of the Figure 4.11 above.

When spatial frequency is determined analytically using the two-dimensional Fourier transform, the result is plotted in the manner of Figure 4.11, where low vertical frequencies – low *y* values – are at the bottom.

When spatial frequency is computed numerically using discrete transforms, such as the 2-D *discrete Fourier transform* (DFT), the *fast Fourier transform* (FFT), or the *discrete cosine transform* (DCT), the result is usually presented in a matrix, where low vertical frequencies are at the top.

I explain the one-dimensional frequency spectrum of NTSC on page 69. There I use an example where every scan line contains a sinewave of frequency f_T, where the phase of the sinewave inverts in alternate lines. In the horizontal domain, all of the power in this signal is located at f_T. In the vertical domain, the situation resembles *Sampling at exactly 0.5fS*, on page 45, where the sample instants correspond to the scan line pitch. All of the power in this signal is contained at a spatial frequency corresponding to the vertical Nyquist frequency.

The vertical spatial frequency that corresponds to half the sampling rate depends on the number of picture lines in the image. If there are 8 picture lines, then 8 vertical samples occupy the height of the picture, and the Nyquist frequency corresponds to 4 C/PH.

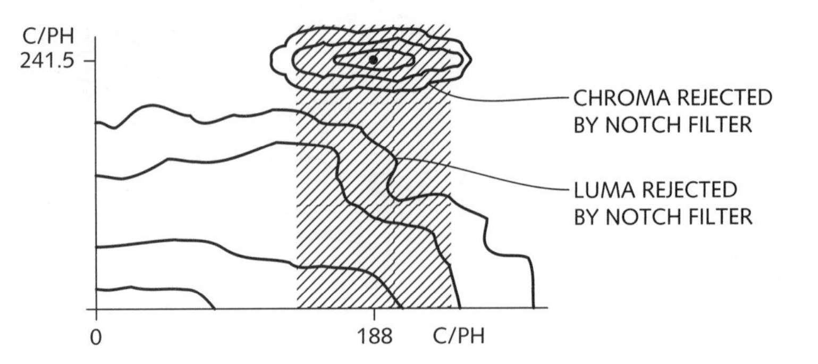

Figure 4.12 **Notch filter in the spatial frequency domain.**

NTSC subcarrier has exactly 227.5 cycles per total line; 188 cycles lie across the width of the picture. Considered in the horizontal spatial domain, subcarrier is located at 188 C/PW.

On page 45, I explained that a signal at $0.5f_S$, with arbitrary phase and amplitude, cannot be sampled without ambiguity. But the amplitude of subcarrier is fixed, and its phase is determined by *burst*, which is undisturbed by picture information. Here, the usual restriction of sampling theory doesn't apply.

The NTSC subcarrier frequency is an odd multiple of half the line rate: its phase inverts at line rate, so its power is located a vertical spatial frequency corresponding to half the scan line pitch. NTSC has 483 picture lines, so subcarrier is located at a vertical frequency of 241.5 C/PH. In Figure 4.11, subcarrier is located near the upper right corner of the plot, at coordinates (188, 241.5).

Viewed in the spatial frequency domain, quadrature modulation causes the chroma signal to be concentrated around the subcarrier. Variations in the color content of the image cause modulated chroma to spread out in spatial frequency around subcarrier, in the same manner that luma spreads from (0, 0). The chroma region is indicated in Figure 4.11.

A notch filter operates only in the horizontal domain. Figure 4.12 above indicates the spatial frequencies that are rejected by a chroma notch filter. Although chroma is rejected, a significant amount of luma is eliminated as well. Better separation can be accomplished by considering vertical and spatial properties.

In *Comb filtering*, on page 68, I explained how to construct a filter that operates in the vertical dimen-

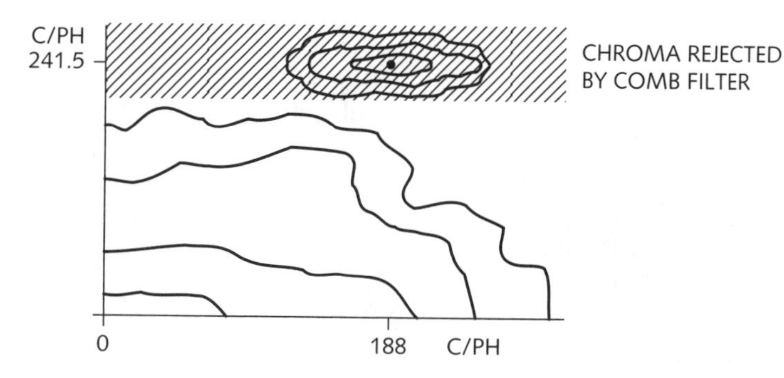

Figure 4.13 **Comb filter in the spatial frequency domain.**

sion. Figure 4.8, on page 69, graphs the response of a comb filter in the vertical domain having coefficients (1, 1). Figure 4.13 above shows the region of spatial frequency that is rejected by this two-line comb filter. It is clear from this graph that luma detail in the range of subcarrier frequencies is retained.

$$\begin{bmatrix} 1 & 2 & 1 \\ 2 & 4 & 2 \\ 1 & 2 & 1 \end{bmatrix}$$

Placing a (1, 2, 1) horizontal lowpass filter in tandem with a (1, 2, 1) vertical lowpass filter is equivalent to computing a weighted sum of spatial samples using the weights indicated in the matrix in the margin. This is an example of a *spatial filter*. This particular spatial filter can be implemented as horizontal and vertical filters in tandem. Many spatial filters are *inseparable*; their computation must take place directly in the two-dimensional spatial domain, and cannot be achieved through cascaded horizontal and vertical filters.

The concept of two-dimensional spatial frequency can be extended to three-dimensional spatiotemporal frequency. Consider the luma component of Figure 4.13 above to be clustered at zero temporal frequency, and imagine a third, temporal axis extending into the page. In addition to inverting line by line, NTSC subcarrier inverts frame by frame. This causes subcarrier to be located at 30 Hz in the third coordinate, and causes chroma to cluster around this point in spatiotemporal frequency. A *frame comb* filter uses a frame of memory to operate along this axis, as sketched in Figure 4.14 in the margin: Summing the indicated samples rejects chroma.

Figure 4.14 **Frame comb.**

Image sampling in computing

Figure 4.15 **Bitmapped graphic image, detail.**

Figure 4.15, in the margin, is taken from a bitmapped graphic image. Each sample is either black or white. The element with horizontal "stripes" is part of a window's titlebar. The checkerboard "desktop pattern" is intended to approximate gray. The image data in this example has been mapped precisely onto the sampling grid of the imaging system. But an image captured from a real scene cannot be guaranteed to have its elements aligned with the raster. Video needs to represent arbitrary images, not just images whose elements are designed to have an intimate relationship with the sampling structure.

An image cannot be accurately digitized unless the continuous information was subjected to a presampling (antialiasing) filter. A suitable filter prohibits image content such as the titlebar and the desktop pattern in Figure 4.15. In a video camera, a spatial presampling filter must be implemented either in the optical path, or within the target of the tube or the photosensor, so as to act prior to conversion of the image signal to electronic form. This is the case both for a "tube" camera, such as a vidicon or a plumbicon, or a CCD camera having discrete photosensor devices.

R. J. Clarke, *Transform Coding of Images*. Boston: Academic Press, 1985.

Any practical image sensor acquires information from a finite region of the image plane. The size and shape of the *sampling aperture* of the image sensor influences the nature of the image signal it originates. Sampling theory can be used to analyze the effects on the image of the sampling aperture.

Image reconstruction

Figure 4.16 opposite shows two different ways to reconstruct the same image data. On the left, the image is reconstructed using a boxcar function. Each pixel is uniformly shaded across its extent. This is also known as *sample-and-hold*, *zero-order hold*, or *nearest-neighbor* reconstruction. On the right, the image is reconstructed using a Gaussian function.

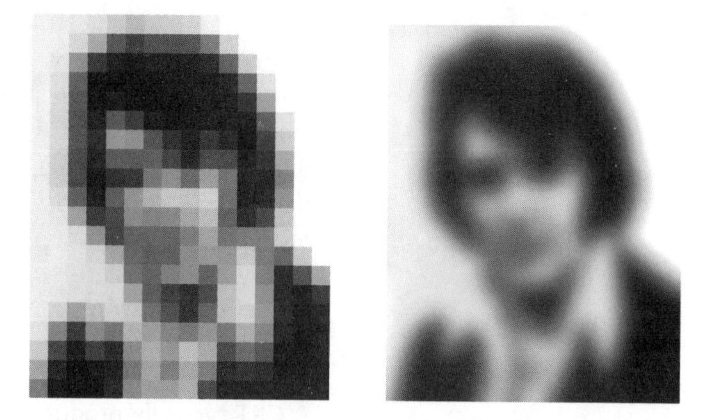

Figure 4.16 **Image reconstruction using a boxcar filter.** At the left is an image of 16×20 pixels that has been reconstructed using a boxcar (*sample-and-hold, zero-order hold,* or *nearest-neighbor*) function. The image is not recognizable. At the right is exactly the same image data, but reconstructed by a Gaussian function. The reconstructed image is very blurry but recognizable.

Each image is 4 cm (1.6 inches) wide, and has 16×20 pixels. At a reading distance of 40 cm (16 inches), a pixel subtends 0.4 degree. Visual acuity is near its maximum at this angular subtense. When the image is reconstructed with a boxcar, the pixel structure of the image is so highly visible that perception of the pixel structure overwhelms the perception of the image itself. There is no suppression of the sampling frequency of the image data, and poor image quality results. When reconstructed by a Gaussian function, the image is blurry but recognizable.

At a viewing distance of 10 m (33 feet), a pixel subtends a minute of arc ($1/60$ degree). At this distance, both images are apparently identical.

The problem with the image reconstructed by a boxcar filter is that a large component at the sampling frequency remains after reconstruction. This frequency component needs to be greatly attenuated so as to minimize the visibility of the sample structure.

You might think that if a reconstruction filter attenuates frequencies at and above the sampling rate, its job is done. But a sampled representation is inherently restricted to frequency components not only less than the sample rate, but less than *half* the sample rate. For accurate, alias-free reconstruction, it is necessary to attenuate frequencies at and above the Nyquist frequency, half the sample rate.

Spot size

Resolution is limited by any mechanism that spreads signal power from one picture element to another. If a lens or display exhibits blurring over the scale of the dimensions of a sample, then resolution will be decreased. *Modulation transfer function* (MTF) characterizes a subsystem or a whole optical system. MTF is simply a one-dimensional plot of horizontal or vertical spatial frequency response, or perhaps just a single point quoted from this graph.

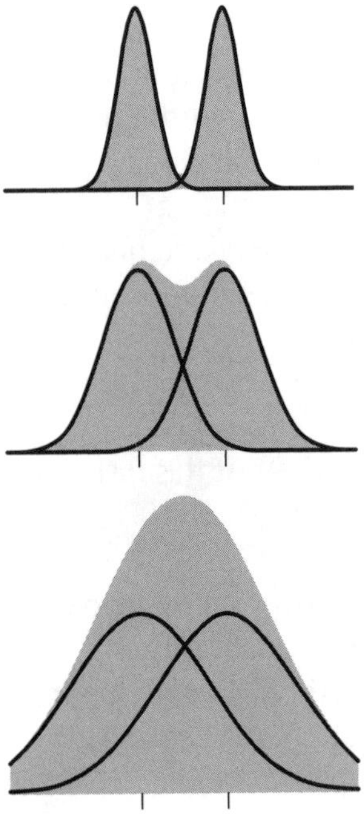

A CRT typically produces a beam (or *spot*) whose intensity distribution resembles a two-dimensional Gaussian function. If there is a gap in the sum of the intensity distributions produced by adjacent pixels, then the display's scan line or pixel structure is likely to be visible. This is the situation in the top sketch in the margin: The solid lines indicate the distribution of intensity across two spots, and the shaded area indicates their sum. A typical CRT has a spot size (at its half-power points) similar to the scan-line pitch, as shown in the middle sketch. Such a large spot size imposes a limit on the resolution that can be achieved at the face of the display, but a visible gap between scan lines is avoided. In the bottom sketch, the half-power point of the spot is twice the scan line pitch; with a spot this large a great deal of potential resolution is lost.

Figure 4.17 **Spot size.**

An image acquisition or display device can be analyzed using filter theory. Theory predicts that resolution can be maximized by using a spot with a sinc distribution. But a sinc function involves negative excursions, so a display employing sinc cannot be realized. A boxcar distribution across each pixel of a sensor or display will cause aliasing at high spatial frequencies, as expected from the sinc frequency response of the boxcar weighting function. If some external mechanism – such as a lens, or an optical filter – is available to attenuate high spatial frequencies, then a boxcar function might be suitable. In the absence of explicit measures to reduce aliasing, a Gaussian can achieve reasonably high resolution while reducing the visibility of the pixel (or scan line) structure and attenuating potential alias components.

Transition samples

At the start of picture information on a scan line, if the video signal immediately assumes a value greatly different from blanking, an artifact called *ringing* results when that rapid transition is processed through an analog or digital filter. A similar circumstance arises at the end of picture on a scan line. In studio quality video, it is important to have the signal build to full amplitude, or decay to blanking level, over a *blanking transition* of a few hundred nanoseconds in the analog domain, or several digital samples. The transition samples should represent a raised cosine transition.

Video standards call for a number of *active* samples – *samples per active line*, S/AL – sufficient to encompass not only the width of the picture, but also the blanking transition samples. In studio quality video, the first and last active samples on a line are at blanking level or very close to it.

Picture center and width

The width of a video picture is measured at the 50 percent point of an entirely white picture (*flat field*). Due to the blanking transition samples mentioned a moment ago, there are somewhat fewer samples across the width of the picture (S/PW) than there are active samples. In component digital video, the horizontal center of the picture is defined to lie midway between two luma samples, and the vertical center of the picture is defined to lie midway between two lines.

In component digital video, the sampling structure aligns with 0_H: If a sample were taken at the horizontal datum (0_H) instant, its value would be the 50 percent point of analog sync. In composite $4f_{SC}$ digital video, a more complicated situation will be detailed in Chapter 14 on page 221, and Chapter 17 on page 241.

Luminance and lightness 5

Vision perceives light having wavelengths in the range of about 400 nm to 700 nm. In *Color science for video*, on page 115, I will describe how power distributions in this region are perceived as colors. This chapter concerns the analysis and reproduction of what is loosely called *brightness*.

First, I will describe *luminance*, denoted *Y*. Luminance is a *linear-light* quantity, directly proportional to physical intensity weighted by the spectral sensitivity of human lightness perception. Luminance can be computed as a properly weighted sum of red, green, and blue linear *tristimulus* components.

In video, we do not compute luminance quite according to the principles of color science. Instead, we compute an approximation *luma*, denoted Y', as a weighted sum of nonlinear (gamma-corrected) R', G', and B' components. Video *luma* is only loosely related to true (CIE) luminance, as I will outline in this chapter and detail in *Luma and color differences*, on page 155. Nonlinear coding of brightness information is essential to maximize the perceptual performance of an image coding system. I will introduce the quantity L^*, defined by the CIE to represent lightness perception. In the chapter *Gamma*, on page 91, I will detail the nonlinear coding of intensity in video.

Radiance, intensity

D. Allan Roberts, "A Guide to Speaking the Language of Radiometry and Photometry," in *Photonics Design and Applications Handbook*, 1994 edition, vol. 3. Pittsfield, MA: Laurin Publications, pages H-70 to H-73.

Radiance – or loosely speaking, *intensity* – refers to power flow in a specified region of the electromagnetic spectrum. Image science is concerned with radiation from, or incident on, a surface. It is measured with an instrument called a *radiometer*, and is what I call a *linear-light* measure, expressed in units such as watts per square meter.

The voltages presented to a CRT monitor control the intensities of the color components, but in a nonlinear manner. CRT voltages are not proportional to intensity, as I will detail in *Gamma*, on page 91.

Luminance

The *Commission Internationale de L'Éclairage* (CIE, or International Commission on Illumination) is the international body responsible for standards in the area of color perception. *Brightness* is defined by the CIE as *the attribute of a visual sensation according to which an area appears to exhibit more or less light*. Brightness is a subjective quantity.

The CIE has defined an objective quantity related to brightness. *Luminance* is defined as radiant power weighted by the spectral sensitivity function – the sensitivity to power at different wavelengths – that is characteristic of vision. The *luminous efficiency* of the CIIE Standard Observer is shown in Figure 5.1 opposite. It is defined numerically, is everywhere positive, and peaks at about 555 nm. When a spectral power distribution (SPD) is integrated using this weighting function, the result is *CIE luminance*, denoted *Y*.
In continuous terms, luminance is an integral over the spectrum. In discrete terms, it is a dot product. The magnitude of luminance is proportional to physical power. In that sense it is like intensity. But the spectral composition of luminance is related to the lightness sensitivity of human vision.

As you will see in *Color science for video*, on page 115, the *luminous efficiency function* is also known as the $\bar{y}$ *color matching function (CMF)*. It is also denoted V(λ), pronounced *vee lambda*.

Strictly speaking, luminance should be expressed in units such as candelas per meter squared ($cd{\cdot}m^{-2}$), but

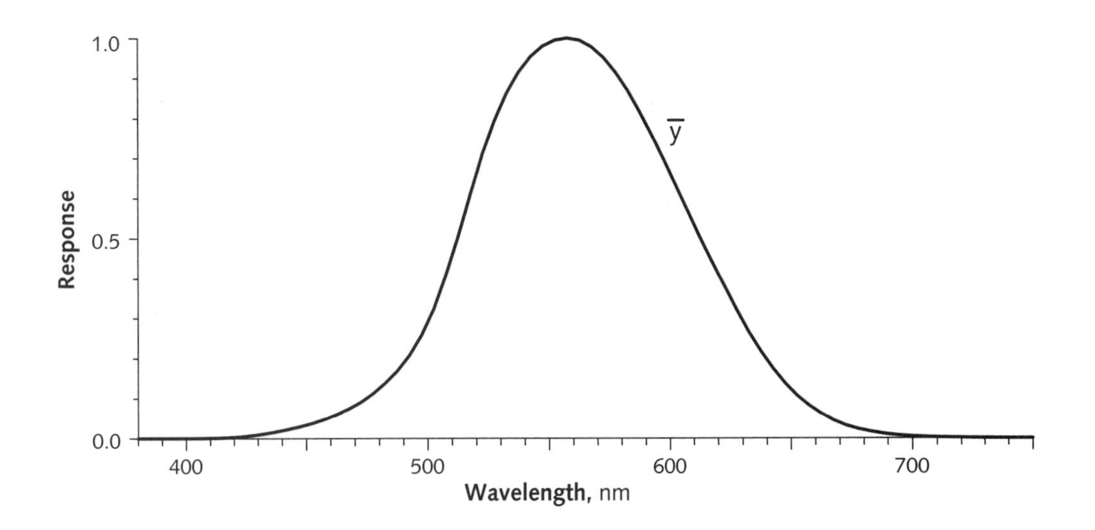

Figure 5.1 **The CIE luminous efficiency function.** A monochrome scanner or camera must have this spectral response in order to correctly reproduce perceived lightness. The function peaks at about 555 nm; a source whose spectrum is concentrated here will be perceived as a bright yellow-green.

SMPTE RP 71-1977, *Setting Chromaticity and Luminance of White for Color Television Monitors Using Shadow-Mask Picture Tubes.*

in practice luminance is often normalized to 1 or 100 units relative to a specified or implied reference *white*. SMPTE standardized a white reference for a studio broadcast monitor of 103 $cd \cdot m^{-2}$.

We intuitively associate pure luminance with gray, but a spectral power distribution having the graph of Figure 5.1 will *not* appear neutral gray! In fact, an SPD of that shape would appear distinctly green. This fact is of little consequence here, but in Chapter 7, *Color science for video*, it will become very important to distinguish analysis functions – there called *color matching functions*, or CMFs – from spectral power distributions. The luminous efficiency function takes the role of an analysis function, not an SPD.

Luminance from red, green, and blue

If the luminance of an image is to be determined by a scanner or camera having a single spectral filter, then the spectral response curve of the scanner's filter must – in theory, at least – correspond to the luminous efficiency function of Figure 5.1. However, luminance can also be computed as a weighted sum of red, green, and blue components.

If three sources appear red, green, and blue, and have the same radiant power in the visible spectrum, then the green will appear the brightest of the three because the luminous efficiency function peaks in the green region of the spectrum. The red will appear less bright, and the blue will be the darkest of the three. As a consequence of the luminous efficiency function, all saturated blue colors are quite dark and all saturated yellows are quite light.

To compute luminance using $\frac{R+G+B}{3}$ is at odds with the properties of vision.

If luminance is computed from red, green, and blue, the coefficients will be a function of the particular red, green, and blue spectral weighting functions employed; but the green coefficient will be quite large, the red will have an intermediate value, and the blue coefficient will be the smallest of the three. The relationship between luminance and spectral sensitivity is complex, but luminance comprises roughly 7 percent power from blue regions of the spectrum, 72 percent from green, and 21 percent from red.

The coefficients that correspond to contemporary CRT displays are standardized in Rec. ITU-R BT.709. These weights are appropriate to compute true CIE luminance from linear red, green, and blue for modern cameras and modern displays:

Eq 5.1

$$Y_{709} = 0.2125\,R + 0.7154\,G + 0.0721\,B$$

I will describe details of this calculation in *Color science for video*, on page 115.

Blue has a small contribution to the brightness sensation. However, human vision has extraordinarily good color discrimination capability in blue hues. If you give blue fewer bits than red or green, blue areas of your images will exhibit contouring artifacts.

Adaptation

Figure 5.2

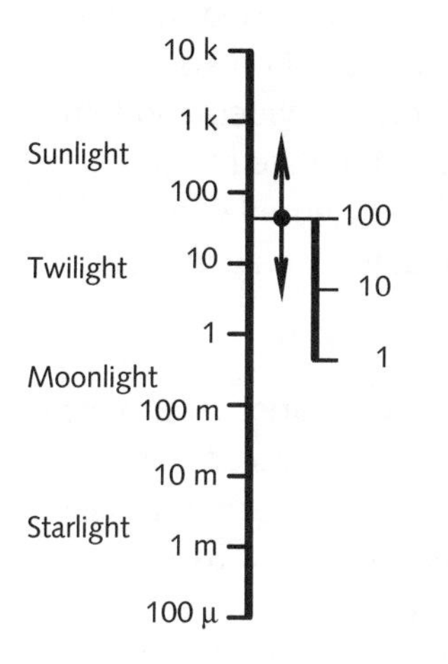

Human vision adapts over a remarkably wide range of intensity levels – about seven decades of dynamic range in total. For about two decades at the bottom end of the intensity range, the retinal photoreceptor cells called *rods*, are employed. Since there is only one type of rod cell, what is loosely called *night vision* cannot discern colors. About one decade of adaptation is effected by the iris; the remainder is due to a photochemical process that involves the *visual pigment* substance contained in photoreceptor cells.

Adaptation is controlled by total retinal illumination. *Dark adaptation* to a lower intensity is slow: It can take many minutes to adapt from a bright sunlit day to the dark ambient of a cinema. Adaptation toward higher intensity is more rapid but can be painful, as you may have experienced when walking out of the cinema back into daylight. Since adaptation is controlled by total retinal illumination, your adaptation state is closely related to the intensity of "white" in your field of view. Effects of adaptation must be carefully considered in order to accurately determine the lightness sensitivity of vision.

Lightness sensitivity

At a particular state of adaptation, human vision can distinguish different luminance levels down to about 1 percent of what is sometimes called "peak white." In other words, our ability to distinguish luminance differences extends over a range of luminances of about 100:1. Loosely speaking, intensities less than about 1 percent of peak white appear simply "black": Different luminances below this level cannot be distinguished.

Contrast ratio is defined as the ratio of luminance between the lightest and darkest elements of a scene. Contrast ratio is a major determinant of perceived picture quality, so much so that an image reproduced with a high contrast ratio may be judged sharper than another image that has higher measured spatial resolution. In a practical imaging system, many factors conspire to increase the luminance of blacks, and thereby lessen the contrast ratio and impair perceived picture quality.

During the course of the day we are exposed to a wide range of illumination levels, and adaptation adjusts accordingly. But viewing conditions for video and film tend to be fairly well controlled. An image reproduction system is almost always concerned with viewing at a known adaptation state, so a contrast ratio of about 100:1 is adequate.

Within the two-decade range of luminance over which vision can distinguish luminance levels, vision has a certain discrimination threshold. For a reason that will become clear in a moment, it is convenient to express the discrimination capability in terms of *contrast sensitivity*, which is the ratio of luminances between two adjacent patches of similar luminance.

Figure 5.3 below shows the pattern presented to an observer in an experiment to determine the contrast sensitivity of human vision. Most of the observer's field of vision is filled by a *surround* luminance level, L_0, which fixes the observer's state of adaptation. In the central area of the field of vision is placed two adjacent patches having slightly different luminance levels, L and $L + \Delta L$. The experimenter presents stimuli having a

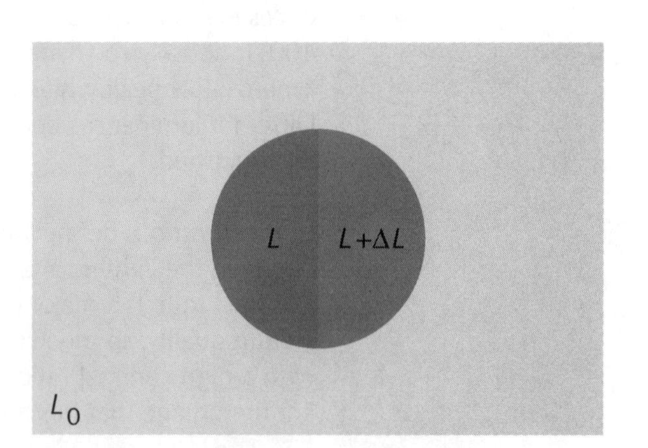

Figure 5.3 **Contrast sensitivity test pattern.** This diagram shows the pattern presented to an observer in an experiment to determine the contrast sensitivity of human vision. The experimenter adjusts ΔL, and the observer is asked to report when he detects a difference in brightness between the two halves of the patch.

L_0: Adaptation (Surround) Luminance

L: Test Luminance

ΔL: Minimum detectable difference

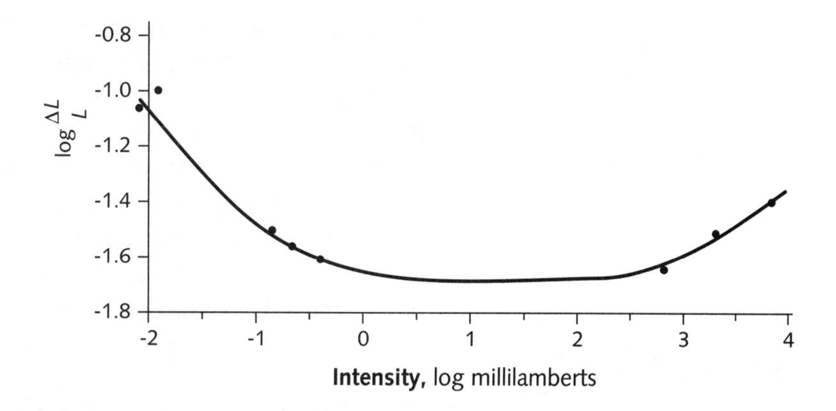

Figure 5.4 **Contrast sensitivity.** This graph is redrawn, with permission, from Figure 3.4 of Schreiber's *Fundamentals of Electronic Imaging Systems.* Over a range of intensities of about 300:1, the discrimination threshold of vision is approximately a constant ratio of luminance. The flat portion of the curve shows that the perceptual response to intensity – termed *lightness* – is approximately logarithmic.

wide range of test values with respect to the surround, that is, a wide range of L/L_0 values. At each test luminance, the experimenter presents to the observer a range of luminance increments with respect to the test stimulus, that is, a range of $\Delta L/L$ values.

William F. Schreiber, *Fundamentals of Electronic Imaging Systems,* Second Edition. Berlin: Springer-Verlag, 1991.

When this experiment is conducted, the relationship graphed in Figure 5.4 above is found: Plotting $\log(\Delta L / L)$ as a function of $\log(L)$ reveals an interval of more than two decades of luminance over which the discrimination capability of vision is about 1 percent of the test luminance level. This leads to the conclusion that – for *threshold* discrimination of two adjacent patches of nearly identical luminance – the discrimination capability is very nearly logarithmic.

$$\frac{\log 100}{\log 1.01} \approx 463$$

Many more than this number of codes is needed to achieve similar performance in a linear-light system. See page 95.

The contrast sensitivity function begins to answer this question: What is the minimum number of discrete codes required to represent luminance over a particular range? In other words, what intensity codes can be thrown away without the observer noticing? If codes are placed at exactly 1-percent intervals over a 100:1 range, about 463 codes are required: That number of codes is required to cover a contrast ratio of 100:1, at a particular adaptation state, at a discrimination threshold of 1 percent.

The logarithmic relationship relates to contrast sensitivity *at threshold*: We are measuring the ability of the visual system to discriminate between two nearly identical luminances. Over a wider range of luminance, strict adherence to logarithmic coding is not justified for perceptual reasons: The discrimination capability of vision degrades for very dark shades of gray, below several percent of peak white. In practice, power functions are used instead of logarithmic functions.

Lightness, CIE L*

The CIE standardized the L^* function to approximate the *lightness* response of human vision. *L-star* is a power function of luminance, modified by the introduction of a linear segment near black. In the following equations, Y is CIE luminance (proportional to intensity), and Y_n is the luminance of reference white, usually normalized to either 1.0 or 100:

Eq 5.2

$$L^* = \begin{cases} 903.3\,\dfrac{Y}{Y_n}, & \dfrac{Y}{Y_n} \le 0.008856 \\ 116\left(\dfrac{Y}{Y_n}\right)^{\frac{1}{3}} - 16, & 0.008856 < \dfrac{Y}{Y_n} \end{cases}$$

L^* has a range of 0 to 100; a *delta L-star* of unity is taken to be roughly the threshold of discrimination. The linear segment is unimportant for practical purposes; but if you don't use it, make sure that you limit L^* at zero. If you normalize luminance to reference white of unity, then you need not compute the quotient.

On page 17, I explained perceptual uniformity using the example of the volume control on a radio. CIE L^* represents a perceptually uniform measure of luminance, or loosely, intensity.

This standard function relates linear-light luminance to perceived lightness. The CIE defines L^* as one component of a *uniform color space*. The term *perceptually linear* is not appropriate: Since we cannot directly measure the quantity in question, we cannot ascribe to it the properties of mathematical linearity. There are other functions – such as Munsell *Value* – that specify alternate lightness scales, but the CIE L^* function is widely used and internationally standardized.

Roughly speaking, perceived lightness is the cube root of luminance. You will see in *Gamma*, on page 91, that video imposes a transfer function similar to L^*, so that the image coding exhibits perceptual uniformity.

Linear and nonlinear processing

Linear-light coding is necessary to simulate the physical world. To produce a numerical simulation of a lens performing a Fourier transform, you should use linear coding. If you want to compare your model with the transformed image obtained from a real lens by a video camera, you will have to remove the nonlinear gamma correction that was imposed by the camera, to convert the image data back into its linear-light representation.

If your computation involves human perception, a nonlinear representation may be required. If you perform a discrete cosine transform (DCT) on image data as the first step in image compression, as in JPEG, then you ought to use nonlinear coding that exhibits perceptual uniformity, because you wish to minimize the perceptibility of the errors that will be introduced during quantization.

The literature of image processing rarely discriminates between linear and nonlinear coding. In the JPEG standard there is no mention of transfer function, but nonlinear (video-like) coding is implicit: Unacceptable results are obtained when JPEG is applied to linear-light data. Computer graphic standards, such as PHIGS and CGM, make no mention of transfer function, but linear-light coding is implicit. These discrepancies make it difficult to exchange high quality imagery in computing.

When you ask a video engineer if his system is linear, he will say Of course – referring to linear voltage. If you ask an optical engineer if her system is linear, she will say Of course – referring to linear intensity. But when a nonlinear transform lies between the two systems, a linear transformation performed in one domain is not linear in the other.

Gamma 6

In photography, video, and computer graphics, the *gamma* symbol, γ, represents a numerical parameter that describes the nonlinearity of intensity reproduction. Gamma is a mysterious and confusing subject, because it involves concepts from four disciplines: physics, perception, photography, and video. This chapter explains how gamma is related to each of these disciplines. Having a good understanding of the theory and practice of *gamma* will enable you to get good results when you create, process, and display pictures.

This chapter focuses on electronic reproduction of images, using video and computer graphics techniques and equipment. I deal mainly with the reproduction of intensity, or, as a photographer would say, *tone scale*. This is one important step to achieving good color reproduction; more detailed information about color can be found in *Color science for video*, on page 115.

A *cathode-ray tube* (CRT) is inherently nonlinear: The intensity of light reproduced at the screen of a CRT monitor is a nonlinear function of its voltage input. From a strictly physical point of view, *gamma correction* can be thought of as the process of compensating for this nonlinearity in order to achieve correct reproduction of intensity.

As explained in *Luminance and lightness*, on page 81, the human perceptual response to intensity is distinctly nonuniform: The *lightness* sensation of vision is roughly a power function of intensity. This characteristic needs to be considered if an image is to be coded so as to minimize the visibility of noise and make effective perceptual use of a limited number of bits per pixel.

Combining these two concepts – one from physics, the other from perception – reveals an amazing coincidence: The nonlinearity of a CRT is remarkably similar to the *inverse* of the lightness sensitivity of human vision. Coding intensity into a gamma-corrected signal makes maximum perceptual use of the channel. If gamma correction were not already necessary for physical reasons at the CRT, we would have to invent it for perceptual reasons.

Photography also involves nonlinear intensity reproduction. Nonlinearity of film is characterized by a parameter *gamma*. As you might suspect, electronics inherited the term from photography! The effect of *gamma* in film concerns the appearance of pictures rather than the accurate reproduction of intensity values. The appearance aspects of *gamma* in film also apply to television and computer displays.

Finally, I will describe how video draws aspects of its handling of *gamma* from all of these areas: knowledge of the CRT from physics, knowledge of the nonuniformity of vision from perception, and knowledge of viewing conditions from photography. I will also discuss additional details of the CRT transfer function that you will need to know if you wish to calibrate a CRT or determine its nonlinearity.

Gamma in physics

The physics of the electron gun of a CRT imposes a relationship between voltage input and light output that a physicist calls a *five-halves power law:* The intensity of light produced at the face of the screen is proportional to the voltage input raised to the power $^5/_2$. Intensity is roughly between the square and cube of the voltage. The numerical value of the exponent of

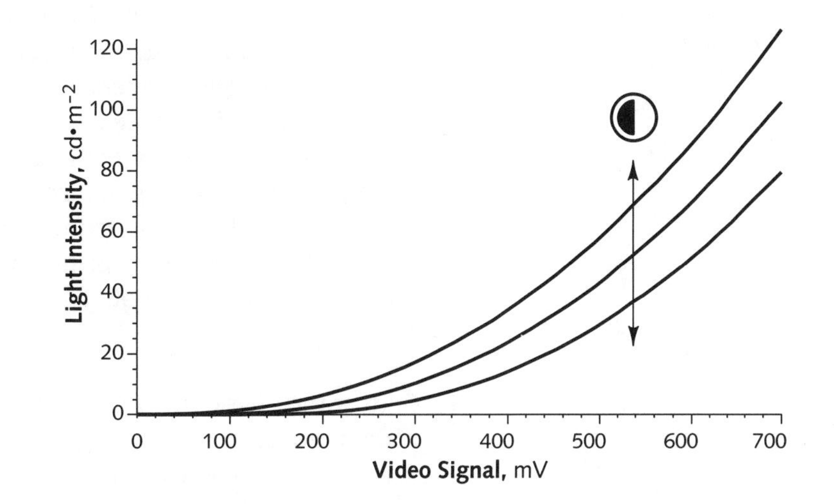

Figure 6.1 **CRT transfer function** involves a nonlinear relationship between video signal and light intensity, here graphed for an actual CRT at three different settings of the *Picture* control. Intensity is approximately proportional to input signal voltage raised to the 2.5 power. The *gamma* of a display system – or more specifically, a CRT – is the numerical value of the exponent of the power function.

the power function is represented by the Greek letter γ (*gamma*). CRT monitors have voltage inputs that reflect this power function. In practice, most CRTs have a numerical value of *gamma* very close to 2.5.

Figure 6.1 above is a sketch of the power function that applies to the single electron gun of a grayscale CRT, or to each of the red, green, and blue electron guns of a color CRT. The functions associated with the three guns of a color CRT are very similar to each other, but not necessarily identical. The function is dictated by the construction of the electron gun; the CRT's phosphor has no significant effect.

Gamma correction involves a power function, which has the form $y = x^a$ (where a is constant). It is sometimes incorrectly claimed to be an exponential function, which has the form $y = a^x$ (where a is constant).

The process of precompensating for this nonlinearity – by computing a voltage signal from an intensity value – is known as *gamma correction*. The function required is approximately a 0.45-power function, whose graph is similar to that of a square root function. In video, gamma correction is accomplished by analog circuits at the camera. In computer graphics, gamma correction is usually accomplished by incorporating the function into a framebuffer's lookup table.

Alan Roberts, "Measurement of display transfer characteristic (gamma, γ)," *EBU Technical Review* 257 (Autumn 1993), 32–40.

The actual value of *gamma* for a particular CRT may range from about 2.3 to 2.6. Practitioners of computer graphics often claim numerical values of *gamma* quite different from 2.5. But the largest source of variation in the nonlinearity of a monitor is caused by careless setting of the *Black Level* (or *Brightness*) control of your monitor. Make sure that this control is adjusted so that black elements in the picture are reproduced correctly before you devote any effort to determining or setting *gamma*.

Getting the physics right is an important first step toward proper treatment of gamma, but it isn't the whole story, as you will see.

The amazing coincidence!

In *Luminance and lightness*, on page 81, I described the nonlinear relationship between luminance and perceived lightness. The previous section described how the nonlinear transfer function of a CRT relates a voltage signal to intensity. Here's the surprising coincidence: The CRT voltage-to-intensity function is very nearly the *inverse* of the luminance-to-lightness relationship of vision. Representing lightness information as a voltage, to be transformed into luminance by a CRT's power function, is very nearly the optimal coding to minimize the perceptibility of noise. CRT voltage is remarkably perceptually uniform.

Suppose you have a luminance value that you wish to communicate to a distant observer through a channel having only 8 bits. Consider a linear light representation, where code zero represents black and code 255 represents white. Code value 100 represents a shade of gray that is approximately at the perceptual threshold: For codes above 100, the ratio of intensity values between adjacent codes is less than 1 percent; and for codes above 100, the ratio of intensity values between adjacent code values is greater than 1 percent.

For luminance values below 100, as the code value decreases toward black, the difference of luminance

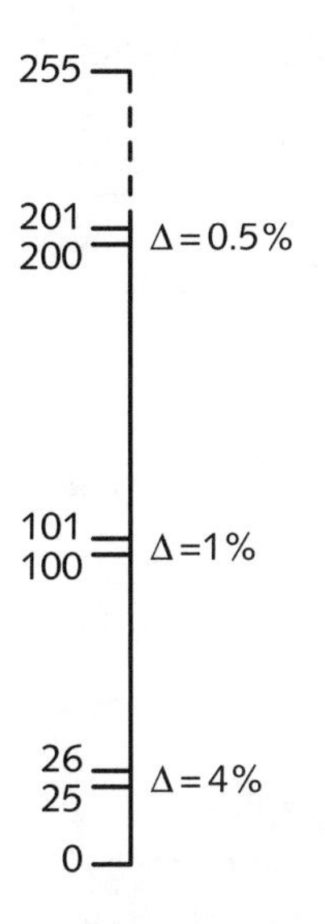

Figure 6.2 **Fixed-point linear-light coding.**

values between adjacent codes becomes increasingly visible: At code 25, the ratio between adjacent codes is 4 percent, which is objectionable to most observers. These errors are especially noticeable in pictures having large areas of smoothly varying shades, where they are known as *contouring* or *banding*.

Luminance codes above 100 suffer no artifacts due to visibility of the jumps between codes. However, as the code value increases toward white, the codes have decreasing perceptual utility. For example, at code 200 the ratio between adjacent codes is 0.5 percent, well below the threshold of visibility. Codes 200 and 201 are visually indistinguishable: Code 201 is perceptually useless and could be discarded without being noticed. This example, sketched in Figure 6.2 in the margin, shows that a linear-luminance representation is a bad choice for an 8-bit channel.

In an image coding system, it is sufficient, for perceptual purposes, to maintain a ratio of luminance values between adjacent codes of about a 1 percent. This can be achieved by coding the signal nonlinearly, as roughly the logarithm of luminance. To the extent that the log function is an accurate model of the contrast sensitivity function, full perceptual use is made of every code.

As mentioned in the previous section, logarithmic coding rests on the assumption that the threshold function can be extended to large luminance ratios. Experiments have shown that this assumption does not hold very well, and coding according to a power law is found to be a better approximation to lightness response than a logarithmic function.

The lightness sensation can be computed as intensity raised to a power of approximately the one-third: Coding a luminance signal to a signal by the use of a power law with an exponent of between $1/3$ and 0.45 has excellent perceptual performance.

S. S. Stevens, *Psychophysics.* New York: John Wiley & Sons, 1975.

Incidentally, other senses behave according to power functions:

Percept	*Physical quantity*	*Power*
Loudness	Sound pressure level	0.67
Saltiness	Sodium chloride concentration	1.4
Smell	Concentration of aromatic molecules	0.6
Heaviness	Mass	1.45

Gamma in film

This section describes gamma in photographic film. I give some background on the photographic process, then explain why physically accurate reproduction of luminance values gives subjectively poor results. Video systems exploit this gem of wisdom from photography: Subjectively better images can be obtained if proper account is taken of viewing conditions.

When film is exposed, light imaged from the scene onto the film causes a chemical change to the emulsion of the film, and forms a *latent image*. Subsequent development causes conversion of the latent image into small grains of metallic silver. This process intrinsically creates a negative image: Where light causes silver to be developed, the developed film absorbs light and appears dark. Color film comprises three layers of emulsion sensitized to different wavelength bands, roughly red, green, and blue. The development process converts silver in these three layers into dyes that act as colored filters to absorb red, green, and blue light.

Film can be characterized by the transfer function that relates exposure to the transmittance of the developed film. When film is exposed in a camera, the exposure value at any point on the film is proportional to the luminance of the corresponding point in the scene, multiplied by the exposure time.

Figure 6.3 **Tone response of color reversal film.** This graph is redrawn, with permission, from Kodak Publication H-1. It shows the S-shaped exposure characteristic of typical color-reversal photographic film. Over the *straight-line* portion of the log-log curve, the density of the developed film is a power function of exposure intensity.

EASTMAN Professional Motion Picture Films, Kodak Publication H-1, Fourth Edition. Rochester, NY: Eastman Kodak Company, 1992. Figure 26.

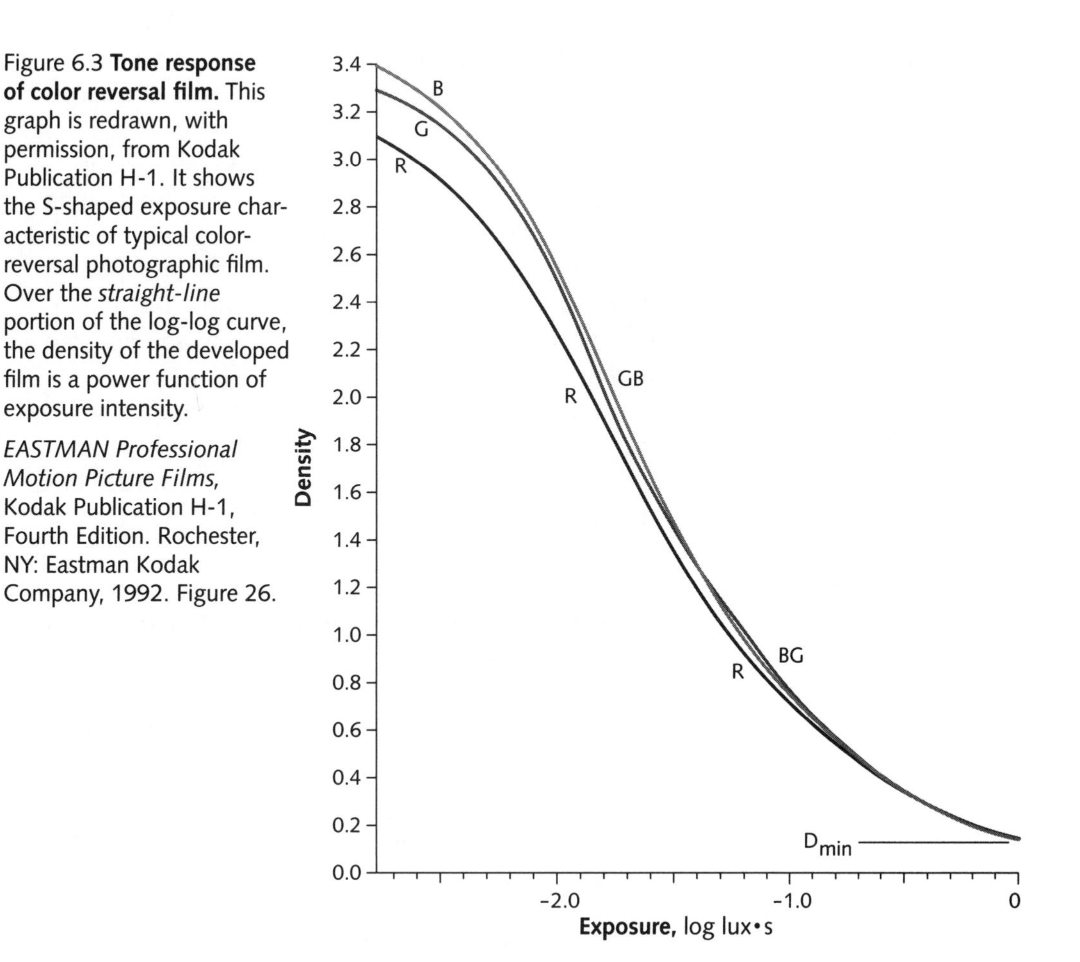

$$D = \log_{10}\left(\frac{P_0}{P_T}\right)$$

D: Density

P_0: Incident Power

P_T: Transmitted Power

See Table 7.5, *Density examples*, on page 152.

Transmittance is defined as the fraction of light incident on the developed film to light absorbed. *Density* is the logarithm of incident power divided by transmitted power. The characteristic of a film is usually shown by plotting *density* as a function of the *logarithm* of *exposure*. This *D*-log *E curve* was first introduced by Hurter and Driffield, so it is also called an *H&D plot*. In terms of the physical quantities of exposure and transmittance, a *D*-log *E* plot is fundamentally in the log-log domain.

A typical film plotted in this way is shown in the plot in Figure 6.3 above. The plot shows an S-shaped curve that compresses blacks, compresses whites, and has a reasonably linear segment in the central portion of the curve. The ubiquitous use of *D*-log *E* curves in film work – and the importance of the linear segment of the curve in determining correct exposure – leads many people to the incorrect conclusion that film has an inherently logarithmic luminance response in terms of physical quantities! But a linear slope on a log-log plot is characteristic of a *power* function, not a logarithmic function: In terms of physical quantities, transmittance of a typical film is a power function of exposure. The slope of the linear segment, in the log-log domain, is the exponent of the power function; in the straight-line region of the film's response curve its numerical value is known as *gamma*.

Since development of film forms a negative image, a second application of the process is necessary to form a positive image; this usually involves making a positive print on paper from a negative on film. In the *reversal* film used in 35 mm slides, developed silver is removed by a bleaching process, then the originally unexposed and undeveloped latent silver remaining in the film is converted to metallic silver to produce a positive image.

This cascaded process is repeated twice in the processing of motion picture film. It is important that the individual power functions at each stage are kept under tight control, both in the design and the processing of the film. To a first approximation, the intent is to obtain roughly *unity* gamma through the entire series of cascaded processes. Individual steps may depart from linearity, as long as approximate linearity is restored at the end of the chain.

Now, here's a surprise. If a film system is designed and processed to produce exactly linear reproduction of intensity, reflection prints look fine. But projected transparencies – slides and movies – look flat, apparently lacking in contrast! The reason for this involves another aspect of human visual perception: the *surround effect*.

Figure 6.4 **Surround effect.** The three gray squares surrounded by white are identical to the three gray squares surrounded by black, but the contrast of the black-surround series appears lower than that of the white-surround series.

Surround effect

As explained in *Adaptation*, on page 85, human vision adapts to an extremely wide range of viewing conditions. One of the mechanisms involved in adaptation increases our sensitivity to small brightness variations when the area of interest is surrounded by bright elements. Intuitively, light from a bright surround can be thought of as spilling or scattering into all areas of our vision, including the area of interest, reducing its apparent contrast. Loosely speaking, the vision system compensates for this effect by "stretching" its contrast range to increase the visibility of dark elements in the presence of a bright *surround*. Conversely, when the region of interest is surrounded by relative darkness, the contrast range of the vision system decreases: Our ability to discern dark elements in the scene decreases. The effect is demonstrated in Figure 6.4 above, from DeMarsh and Giorgianni.

LeRoy E. DeMarsh and Edward J. Giorgianni, "Color Science for Imaging Systems," in *Physics Today*, September 1989, 44–52.

The surround effect has implications for the display of images in dark areas, such as projection of movies in a cinema, projection of 35 mm slides, or viewing of television in your living room. If an image is viewed in a *dark* or *dim surround*, and the intensity of the scene is reproduced with correct physical intensity, the image will appear lacking in contrast.

Film systems are designed to compensate for viewing surround effects. Transparencies (slide) film is intended for viewing in a dark surround. Slide film is designed to have a gamma considerably greater than unity – about 1.5 – so that the contrast range of the scene is expanded upon display. Video signals are coded in a similar manner, taking into account viewing in a dim surround, as I will describe in a moment.

The important conclusion to take from this section is that image coding for the reproduction of pictures for human viewers is not simply concerned with mathematics, physics, chemistry, and electronics. Perceptual considerations play an essential role in successful image systems.

Gamma in video

In a video system, gamma correction is applied at the camera for the dual purposes of coding into perceptually uniform space and precompensating the nonlinearity of the display's CRT. Figure 6.5 opposite summarizes the image reproduction situation for video. Gamma correction is applied at the camera, at the left; the display, at the right, imposes the inverse power function.

Coding into a perceptual domain was important in the early days of television because of the need to minimize the noise introduced by over-the-air transmission. However, the same considerations of noise visibility apply to analog videotape recording, and also to the quantization noise that is introduced at the front end of a digital system when a signal representing intensity is quantized to a limited number of bits. Consequently, it is universal to convey video signals in gamma-corrected form.

As explained in *Gamma in film*, on page 96, it is important for perceptual reasons to "stretch" the contrast ratio of a reproduced image when viewed in a dim surround. The dim surround condition is characteristic of television viewing. In video, the "stretching" is

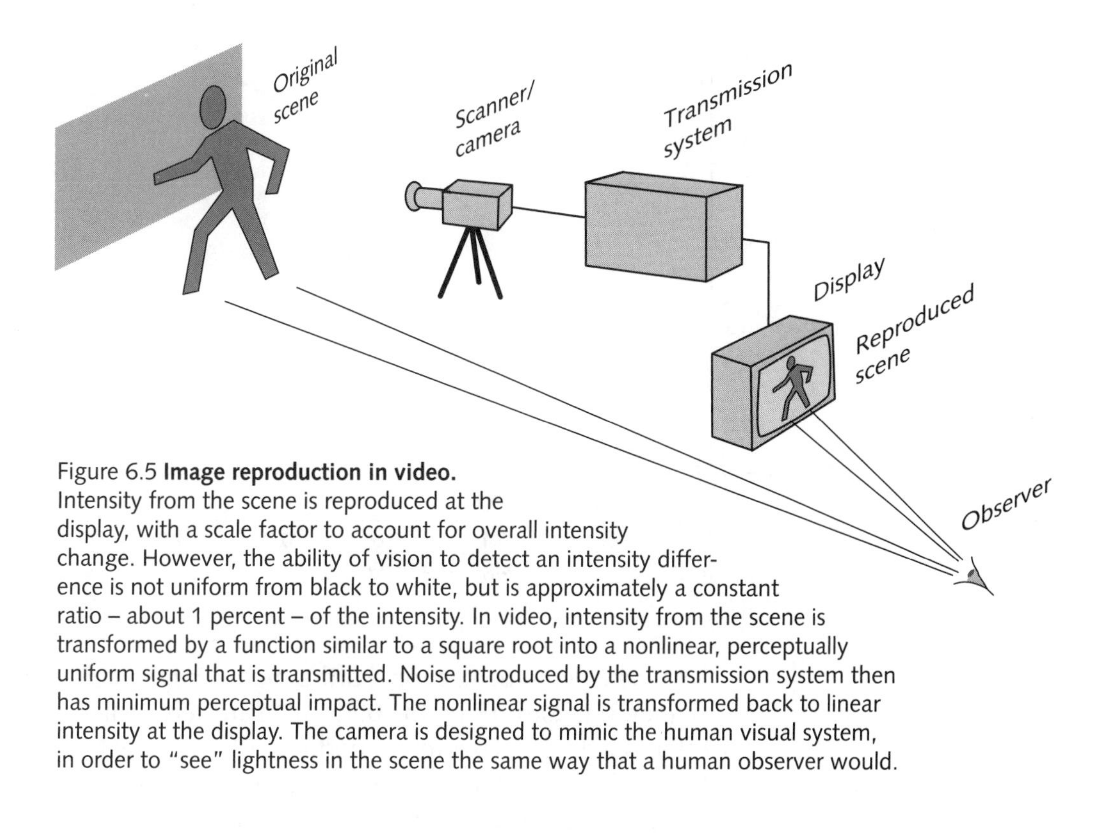

Figure 6.5 **Image reproduction in video.** Intensity from the scene is reproduced at the display, with a scale factor to account for overall intensity change. However, the ability of vision to detect an intensity difference is not uniform from black to white, but is approximately a constant ratio – about 1 percent – of the intensity. In video, intensity from the scene is transformed by a function similar to a square root into a nonlinear, perceptually uniform signal that is transmitted. Noise introduced by the transmission system then has minimum perceptual impact. The nonlinear signal is transformed back to linear intensity at the display. The camera is designed to mimic the human visual system, in order to "see" lightness in the scene the same way that a human observer would.

accomplished at the camera by slightly undercompensating the actual power function of the CRT to obtain an end-to-end power function with an exponent of 1.1 or 1.2. This achieves pictures that are more subjectively pleasing than would be produced by a mathematically correct linear system.

$$0.45 = \frac{1}{2.22\dot{2}}$$

$$\frac{1}{2.2} = 0.4\dot{5}4\dot{5}$$

$$0.45 \times 2.5 \approx 1.13$$

Rec. 709 specifies a power function exponent of 0.45. The product of the 0.45 exponent at the camera and the 2.5 exponent at the display produces the desired end-to-end exponent of about 1.13. An exponent of 0.45 is a good match for both CRTs and for perception. Some video standards have specified an exponent of $^1/_{2.2}$.

Emerging display devices such as liquid crystal displays (LCDs) have nonlinearity different from that of a CRT. But it remains important to use image coding that is

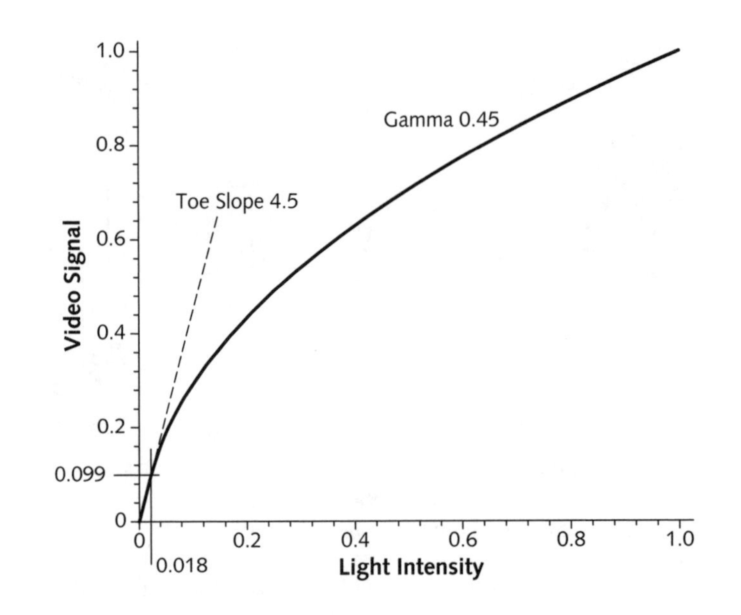

Figure 6.6 **Rec. 709 transfer function.**

well matched to perception. Furthermore, image interchange standards using the 0.45 value are very well established. The economic importance of equipment that is already built around these standards will deter any attempt to establish new standards just because they are better matched to particular devices. We can expect new display devices to incorporate local correction, to adapt between their intrinsic transfer functions and the transfer function that has been standardized for image interchange.

Rec. 709 transfer function

Figure 6.6 above illustrates the transfer function defined by the international Rec. 709 standard for high-definition television (HDTV). It is basically a power function with an exponent of 0.45. Theoretically a pure power function suffices for gamma correction; however, the slope of a pure power function is infinite at zero. In a practical system such as a television camera, in order to minimize noise in the dark regions of the picture it is necessary to limit the slope (gain) of the function near black. Rec. 709 specifies a slope of 4.5 below a tristimulus value of +0.018, and stretches the remainder of the curve to maintain function and tangent continuity at the breakpoint. In this equation the red tristimulus (linear light) component is denoted R, and the resulting gamma-corrected video signal is denoted with a prime symbol, R'_{709}. The computation is identical for the other two components:

Eq 6.1

$$R'_{709} = \begin{cases} 4.5\,R, & R \le 0.018 \\ 1.099\,R^{0.45} - 0.099, & 0.018 < R \end{cases}$$

Standards for conventional 525/59.94 video have historically been very poorly specified. The original NTSC standard called for precorrection assuming a display power function of 2.2. Modern 525/59.94 standards have adopted the Rec. 709 function.

Formal standards for 625/50 video call for precorrection for an assumed power function exponent of 2.8 at the display. This is unrealistically high. In practice the Rec. 709 transfer function works well.

SMPTE 240M transfer function

SMPTE Standard 240M for 1125/60 HDTV was adopted several years before international agreement was achieved on Rec. 709. Virtually all HDTV equipment that has been deployed as I write this uses SMPTE 240M parameters. The 240M parameters are slightly different from those of Rec. 709:

Eq 6.2

$$R'_{240M} = \begin{cases} 4.0\,R, & R \le 0.0228 \\ 1.1115\,R^{0.45} - 0.1115, & 0.0228 < R \end{cases}$$

The difference between the SMPTE 240M and Rec. 709 transfer functions is negligible for real images. It is a shame that international agreement could not have been reached on the SMPTE 240M parameters that were widely implemented at the time the CCIR (now ITU-R) discussions were taking place.

The Rec. 709 values are closely representative of current studio practice, and should be used for all but very unusual conditions.

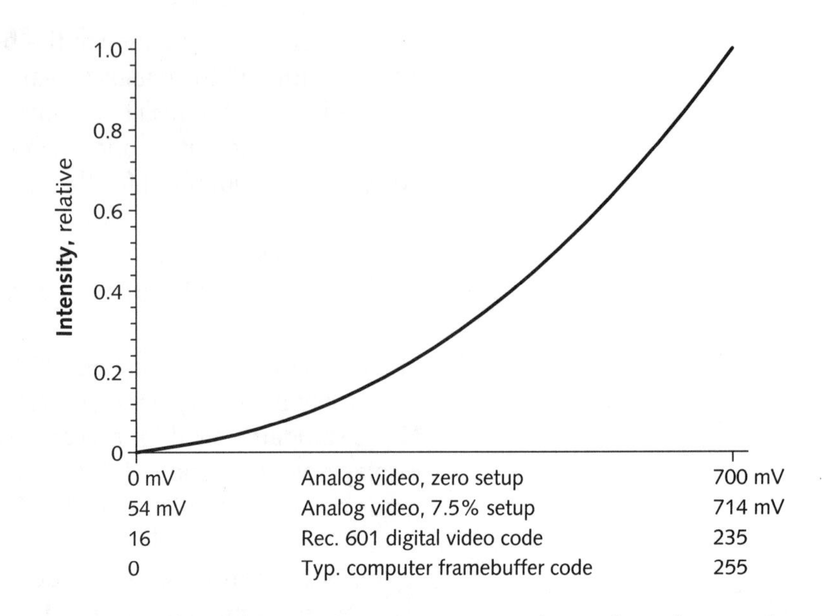

Figure 6.7 **CRT signal levels and intensity.** A video signal may be represented as analog voltage, with zero setup or with 7.5-percent setup. Alternatively, the signal may be represented digitally using Rec. 601 coding from 16 to 235, for studio video, or coding from 0 to 255, used in computer graphics.

CRT transfer function details

This section provides technical information concerning the nonlinearity of a CRT. This section is important if you wish to determine the transfer function of your CRT, to calibrate your monitor, or to understand the electrical voltage interface between a computer framebuffer and a monitor.

Figure 6.7 above illustrates the function that relates signal input to a CRT monitor to the light intensity produced at the face of the screen. The graph characterizes a grayscale monitor, or each of the red, green, and blue components of a color monitor. The *x*-axis of the graph shows the input signal level, from reference black to reference white. The input signal can be presented as a digital code or an analog voltage according to one of several standards. The *y*-axis shows the resulting intensity.

For analog voltage signals, two standards are in use. The range 54 mV to 714 mV is used in video systems that have *7.5-percent setup*, including composite 525/59.94 systems such as NTSC, and computer video systems that conform to the *levels* of the archaic *EIA RS-343-A* standard. Computer framebuffer digital-to-analog converters often have 7.5-percent setup; these almost universally have very loose tolerance of about ±5 percent of full scale on the analog voltage associated with reference black. This induces black level errors, which in turn cause serious errors in the intensity reproduced for black. In the absence of a display calibrator, you must compensate these framebuffer black-level errors by adjusting the *Black Level* (or *Brightness*) control on your monitor. This act effectively marries the monitor to the framebuffer.

The accuracy of black level reproduction is greatly improved in newer analog video standards that have *zero setup*. The voltage range 0 to 700 mV is used in zero-setup standards, including 625/50 video in Europe, and all HDTV standards and proposals.

For the 8-bit digital *RGB* components that are ubiquitous in computing, reference black corresponds to digital code 0, and reference white corresponds to digital code 255. The standard Rec. 601 coding for studio digital video places black at code 16 and white at code 235. Either of these digital coding standards can be used in conjunction with an analog interface having either 7.5-percent setup or zero setup. Coding of imagery with an extended color gamut may place the black and white codes even further inside the coding range, for reasons having to do with color reproduction that are outside the scope of this chapter.

The nonlinearity in the voltage-to-intensity function of a CRT originates with the electrostatic interaction between the cathode and the grid that controls the current of the electron beam. Contrary to popular opinion, the CRT phosphors themselves are quite linear, at least up to an intensity of about eight-tenths of peak white at the onset of saturation.

Knowing that a CRT is intrinsically nonlinear, and that its response is based on a power function, many users have attempted to summarize the nonlinearity of a CRT display in a single numerical parameter using the relationship:

Eq 6.3

$$intensity = voltage^{\gamma}$$

This model shows wide variability in the value of gamma, mainly due to black-level errors that the model cannot accommodate due to its being "pegged" at zero: The model forces zero voltage to map to zero intensity for *any* value of gamma. Black-level errors that displace the transfer function upward can be "fit" only by choosing a gamma value that is much smaller than 2.5. Black-level errors that displace the curve downward – saturating at zero over some portion of low voltages – can get a good "fit" only by having a value of gamma that is much larger than 2.5. In effect, the only way the single gamma parameter can fit a black-level variation is to alter the curvature of the function. The apparent wide variability of gamma under this model has given *gamma* a bad reputation.

A much better model is obtained by fixing the exponent of the power function at 2.5, and using a single parameter to accommodate black-level error:

Eq 6.4

$$intensity = (voltage + \varepsilon)^{2.5}$$

This model fits the observed nonlinearity much better than the variable-gamma model.

William B. Cowan, "An Inexpensive Scheme for Calibration of a Colour Monitor in terms of CIE Standard Coordinates," in *Computer Graphics*, vol. 17, no. 3 (July 1983), 315–321.

If you want to determine the nonlinearity of your monitor, consult the article by Cowan. In addition to describing how to measure the nonlinearity, he describes how to determine other characteristics of your monitor – such as the chromaticity of its white point and its primaries – that are important for accurate color reproduction.

Gamma in computer graphics

Computer graphics software systems generally perform calculations for lighting, shading, depth-cueing, and antialiasing using intensity values that model the physical mixing of light. Intensity values stored in the framebuffer are gamma-corrected by hardware lookup tables on the fly on their way to the display. The power function at the CRT acts on the gamma-corrected signal voltages to reproduce the correct intensity values at the face of the screen. Software systems usually provide a default gamma value and some method to change the default.

The voltage between 0 and 1 required to display a red, green, or blue intensity between 0 and 1 is this:

Eq 6.5

$$signal = intensity^{\left(\frac{1}{\gamma}\right)}$$

In the C language this can be represented as follows:

```
signal = pow((double)intensity,(double)1.0/gamma);
```

In the absence of data regarding the actual gamma value of your monitor, or to encode an image intended for interchange in gamma-corrected form, the recommended value of *gamma* is $1/_{0.45}$ (or about 2.222).

You can construct a gamma-correction lookup table suitable for computer graphics applications, like this:

```
#define SIG_FROM_INTEN(i) \
   ((int)( 255.0 * pow((double)(i) / 255.0, 0.45)))
int sig_from_inten[256], i;
for (i=0; i<256; i++)
   sig_from_inten[i] = SIG_FROM_INTEN(i);
```

Loading this table into the hardware lookup table at the output side of a framebuffer will cause *RGB* intensity values with integer components between 0 and

255 to be gamma-corrected by the hardware as if by the following C code:

```
red_signal = sig_from_inten[r];
green_signal = sig_from_inten[g];
blue_signal = sig_from_inten[b];
```

A lookup table at the output of the framebuffer enables signal representations other than linear-light. If gamma-corrected video signals are loaded into the framebuffer, then a unity ramp is appropriate at the lookup table. This arrangement will maximize perceptual performance.

The availability of a lookup table at the framebuffer makes it possible for software to perform tricks, such as inverting all of the lookup table entries momentarily to flash the screen without modifying any data in the framebuffer. Direct access to framebuffer lookup tables by applications makes it difficult or impossible for system software to avoid annoyances, such as colormap flashing, and to provide features such as accurate color reproduction. To allow the user to make use of these features, applications should access lookup tables in the structured ways that are provided by the graphics system.

Gamma in video, computer graphics, SGI, and Macintosh

Transfer functions in video, computer graphics, Silicon Graphics, and Macintosh are sketched in Figure 6.8 opposite. Video is shown in the top row. Gamma correction is applied at the camera, and signals are maintained in a perceptual domain throughout the system until conversion back to intensity at the CRT.

What are loosely called *JPEG files* use the *JPEG File Interchange Format* (JFIF). Version 1.02 of that specification states that linear-light coding (gamma 1.0) is used. That is seldom the case in practice. Instead, power laws of 0.45, $1/1.8 \approx 0.55$, or $1.7/2.5 \approx 0.68$ are used.

Computer graphics systems generally store intensity values in the framebuffer, and gamma-correct on the fly through hardware lookup tables on the way to the display, as illustrated in the second row.

Silicon Graphics computers, by default, use a lookup table with a 1.7-power function; this is shown in the third row.

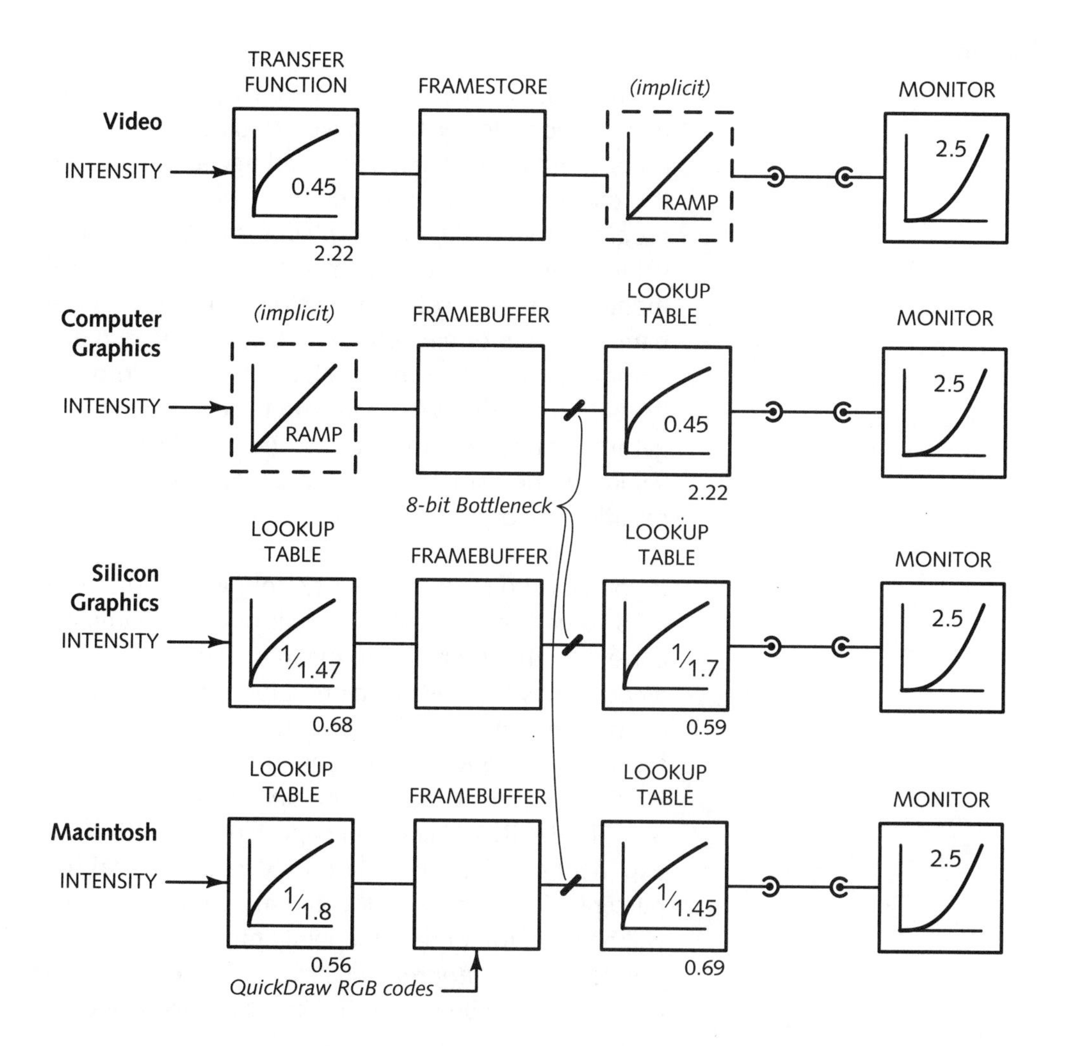

Figure 6.8 **Gamma in video, computer graphics, SGI, and Macintosh.** In a video system, shown in the top row, a transfer function in accordance with vision is applied at the camera. The middle row illustrates computer graphics: Calculations are performed in the linear light domain and gamma correction is applied in a lookup table at the output of the framebuffer. Silicon Graphics computers take a hybrid approach: Part of the correction is accomplished in software, and a $1/1.7$ power function is loaded into the lookup table. The approach used by Macintosh computer sketched in the bottom row.

JFIF files originated on Macintosh ordinarily encode R, G, and B tristimulus (intensity) values raised to the $1/1.8$ power.

Macintosh computers use the approach shown in the bottom row. Part of gamma correction is effected by application software prior to presentation of *RGB* values to the QuickDraw graphics subsystem; the remainder is accomplished in the lookup tables. The dominance of Macintosh computers in graphic arts and prepress has made "gamma 1.8" a *de facto* standard.

Pseudocolor

In *Raster images in computing*, on page 33, I described how pseudocolor systems have lookup tables whose outputs are directly mapped to voltage at the display. It is conventional for a pseudocolor application program to provide, to a graphics system, *RGB* color values that are already gamma-corrected for a typical monitor. A pseudocolor image stored in a file is accompanied by a *colormap* whose *RGB* values incorporate gamma correction. If these values are loaded into a 24-bit framebuffer whose lookup table is arranged to gamma-correct intensity values, the pseudocolor values will be gamma-corrected a second time, resulting in poor image quality.

If you want to recover intensity from gamma-corrected *RGB* values, for example to "back-out" the gamma correction that is implicit in the *RGB* colormap values associated with an 8-bit colormapped image, construct an inverse-gamma table. You can employ a lookup technique as above, building an inverse table INTEN_FROM_SIG using code similar to the SIG_FROM_INTEN code on page 107, but with an exponent of $1/_{0.45}$ instead of 0.45. Be aware that the perceptual uniformity of the gamma-corrected image will be compromised by mapping into the 8-bit intensity domain: *Contouring* – which I will discuss on page page 113 – will be introduced into the darker shades.

Halftoning

Figure 6.9

Continuous-tone (grayscale or color) image data can be reproduced using a process, such as color photography or thermal dye transfer printing, where a continuously variable amount of color material can be deposited at each point in the image. But some reproduction processes – offset lithographic printing and laserprinters, for example – place a fixed density of color at each point in the reproduced image. Grayscale and color images are *halftoned* – or *screened* – in order to be displayed on these devices. Halftoning produces apparent continuous tone by varying the area of small dots in a regular array. Viewed at a sufficient distance, an array of small dots produces the perception of light gray, and an array of large dots produces dark gray.

Halftone dots are usually placed on a regular grid, although *stochastic screening* has recently been introduced, which modulates the spacing of the dots rather than their size. The screening is less visible because the pattern is not spatially correlated.

In order for halftoning to produce a reasonably good impression of continuous tone, the individual dots must not be too evident. If it is known that a reproduction will be viewed at a certain distance, then the number of screen lines per inch can be determined. This measurement can be expressed in terms of the angle subtended by the screen line pitch at the intended viewing distance: If the screen line pitch subtends any more than about one minute of arc at the viewer's retina, screen lines are likely to be visible.

Robert Ulichney, *Digital Halftoning*. Boston: MIT Press, 1988.

Peter Fink, *PostScript Screening: Adobe Accurate Screens*. Mountain View, CA: Adobe Press, 1992.

The standard reference to halftoning algorithms is Ulichney, but he does not detail the nonlinearities found in practical printing systems. For details about screening for color reproduction, consult Fink.

Printing

An image destined for halftone printing conventionally specifies each pixel as *dot percentage in film*. An imagesetter's halftoning machinery generates dots whose areas are proportional to the requested coverage. In principle, *dot percentage in film* is inversely proportional to linear-light reflectance.

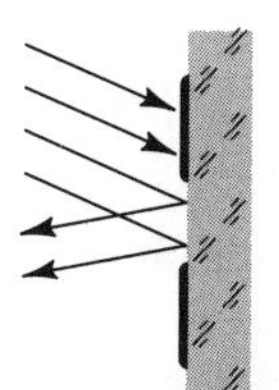

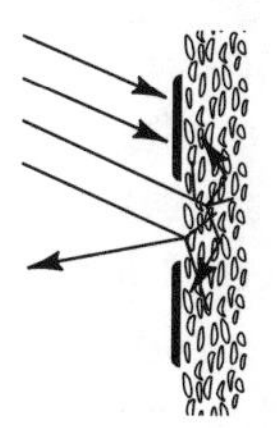

Figure 6.10 **Dot gain mechanism.**

Two phenomena distort the requested dot coverage values. First, printing involves a mechanical smearing of the ink that causes dots to enlarge. Second, optical effects within the bulk of the paper cause more light to be absorbed than would be expected from the surface coverage of the dot alone. These phenomena are collected under the term *dot gain*, which is the percentage by which the light absorption of the printed dots exceeds the requested dot coverage.

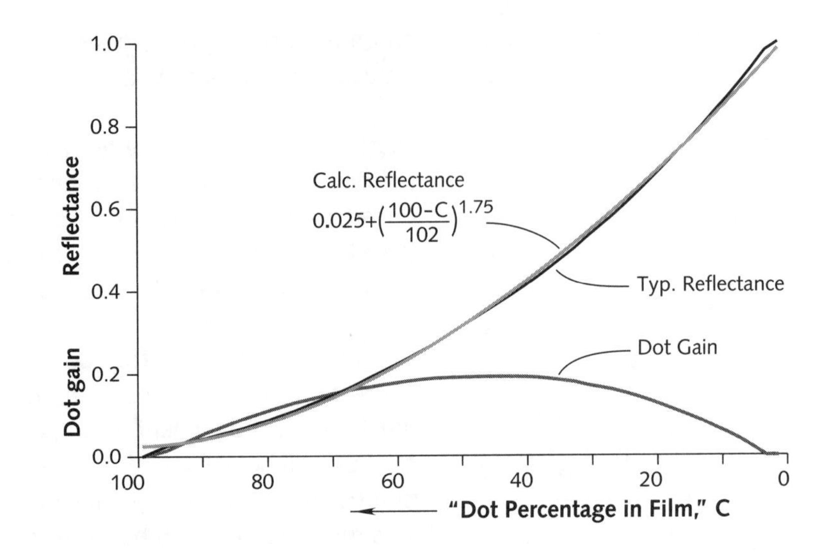

Figure 6.11 **Transfer function in offset printing.** *Dot gain* refers to light absorption in excess of that predicted by ink coverage, or *dot percentage in film*, alone. When expressed as intensity instead of absorption, and as total absorption instead of excess absorption, the standard dot gain characteristic of offset printing reveals a transfer function similar to that of a CRT.

Standard offset printing produces a dot gain at 50 percent of about 22 percent: When 50 percent absorption is requested, 72 percent absorption is obtained. The midtones print darker than requested. This results in a transfer function from code to reflectance that closely resembles the voltage-to-light curve of a CRT. Correction of dot gain is conceptually similar to gamma correction in video: Physical correction of the "defect" in the reproduction process is very well matched to the lightness perception of human vision. Coding an image in terms of dot percentage in film involves coding into a roughly perceptually uniform space. The standard dot gain functions employed in North America and Europe correspond to intensity being reproduced as a power function of the digital code, where the numerical value of the exponent is about 1.75, compared to about 2.2 for video or 3 for CIE L^*. The value used in printing is lower than the optimum for perception, but works well for the low contrast ratio of offset printing.

The Macintosh has a power function that is so similar to that of printing that raw QuickDraw codes sent to an imagesetter produce acceptable results. High-end publishing software allows the user to specify the parameters of dot gain compensation.

I have described the linearity of conventional offset printing. Other halftoned devices have different characteristics, and require different corrections.

Limitations of 8-bit intensity

As mentioned in *Gamma in computer graphics*, on page 107, computer graphics systems that render synthetic imagery usually perform computations in the linear-light (intensity) domain. Graphics accelerators usually perform Gouraud shading in the intensity domain, and store 8-bit intensity components in the framebuffer. Eight-bit intensity representations suffer contouring artifacts, due to the contrast sensitivity threshold of human vision discussed in *Lightness sensitivity*, on page 85. The visibility of contouring is enhanced by a perceptual effect called *Mach bands*; consequently, the artifact is sometimes called *banding*.

In fixed-point intensity coding where black is code zero, code 100 is at the threshold of visibility at 1 percent contrast sensitivity: Code 100 represents the darkest gray that can be reproduced without the increments between adjacent codes being perceptible. I call this value *best gray*. One of the determinants of the quality of an image is the intensity ratio between *brightest white* and *best gray*. In an 8-bit linear-light system, this ratio is a mere 2.5:1. If an image is contained within this contrast ratio, then it will not exhibit banding but the low contrast ratio will cause the image to appear flat. If an image has a contrast ratio substantially larger than 2.5:1, then it is liable to show banding. In 12-bit linear light coding the ratio improves to 82:1, which is adequate for the office but does not approach the quality of a photographic reproduction.

High-end systems for computer generated imagery (CGI) usually do not depend on hardware acceleration. They perform rendering calculations in the intensity domain, perform gamma correction in software, then write gamma-corrected values into the framebuffer. These systems produce rendered imagery without the quantization artifacts of 8-bit intensity coding.

The future of gamma correction

Work is underway to implement facilities in graphics systems to allow device-independent specification of color. Users and applications will be able to specify colors, based on the CIE standards, without concern for gamma correction. When this transition is complete, it will be much easier to obtain color matching across different graphics libraries and different hardware. In the meantime, you can take the following steps:

- Establish good viewing conditions. If you are using a CRT display, you will get better image quality if your overall ambient illumination is reduced.

- Ensure that your monitor's *Black Level* (or *Brightness*) control is set to correctly reproduce black elements on the screen.

- Use gamma-corrected representations of *RGB* values whenever you can. An image coded with gamma correction has good perceptual uniformity, resulting in an image with much higher quality than one coded as 8-bit intensity values.

- When you exchange images either in truecolor or pseudocolor form, code $R'G'B'$ color values using the Rec. 709 gamma value of $^1/_{0.45}$.

- In the absence of reliable information about your monitor, display pictures assuming a monitor gamma value of 2.5. If you view your monitor in a dim surround, use a lower value of about 2.2.

Color science for video 7

Video processing is generally concerned with color represented in three components derived from the scene, usually red, green, and blue, or components computed from these. But accurate color reproduction depends on knowing exactly how the physical spectra of the original scene are transformed into these components, and exactly how the components are transformed to physical spectra at the display. These issues are the subject of this chapter.

Color science establishes the basis for numerical description of color. But classical color science is intended for the *specification* of color, not for image coding. Although an understanding of color science is necessary to achieve good color performance in video, its strict application is unsuitable. This chapter explains the engineering compromises necessary to make practical cameras and practical coding systems.

Once red, green, and blue components of a scene are obtained, *color coding* transforms these components into other forms optimized for processing, recording, and transmission. That topic is discussed in *Component video color coding*, on page 171.

If you are unfamiliar with the term *luminance*, or the symbols *Y* or *Y'*, refer to *Luminance and lightness*, on page 81.

I use the term *luminance* and the symbol *Y* to refer to CIE luminance, a linear-light quantity. I use the term *luma* and the symbol *Y'* to refer to the video signal that conveys luminance information.

Should you wish to skip this chapter, remember two things. First, accurate description of colors expressed in terms of *RGB* coordinates depends on the characterization of the *RGB* primaries. If your system is standardized to use a fixed set of primaries throughout, you may not need to be concerned about this; but if your system uses different primary sets, it is a vital issue.

Second, for high accuracy, color transforms should be computed in the linear-light domain. However, if two primary sets have chromaticity coordinates that are fairly close, reasonably good results can be obtained by computing nonlinear (gamma-corrected) image data.

Fundamentals of vision

Color is the perceptual result of light having wavelengths from 400 nm to 700 nm that is incident upon the retina.

Robert M. Boynton, *Human Color Vision*. New York: Holt, Rinehart and Winston, 1979.

The human retina has three types of color photoreceptor *cone* cells, which respond to incident radiation with somewhat different spectral response curves. A fourth type of photoreceptor cell, the *rod*, is also present in the retina. Rods are effective only at extremely low light levels. Because there is only one type of rod cell, "night vision" cannot perceive color. Image reproduction takes place at light levels sufficiently high that the rod receptors play no role.

Because there are exactly three types of color photoreceptors, three numerical components are necessary and sufficient to describe a color, provided that appropriate spectral weighting functions are used: Color vision is inherently *trichromatic*.

Power distributions exist in the physical world, but color exists only in the eye and the brain. Sir Isaac Newton put it this way:

"Indeed rays, properly expressed, are not coloured."

Color specification

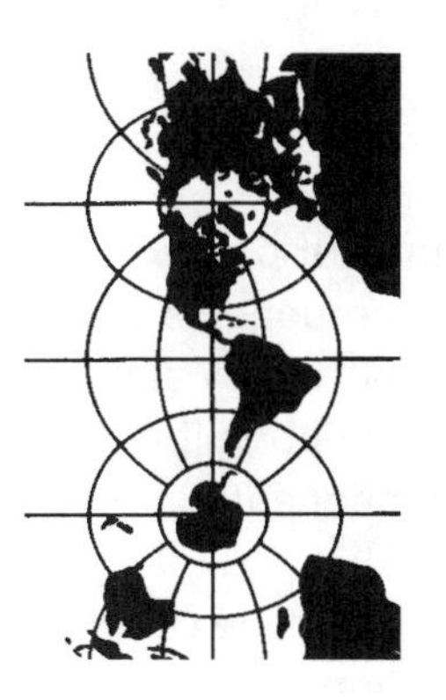

The *Commission Internationale de L'Éclairage* (CIE) has defined several systems to map a *spectral power distribution* (SPD), which I will explain in the following sections, to a *triple* of numerical components that are the mathematical coordinates of color space. These coordinates are analogous to coordinates on a map. Cartographers have different map projections for different functions: Some map projections preserve areas, others show latitudes and longitudes as straight lines. No single map projection fills all the needs of map users. Similarly, no single color system fills all of the needs of color users.

The systems useful for color specification are all based on CIE *XYZ*. I will describe CIE *xyY*, CIE $L^*u^*v^*$, and CIE $L^*a^*b^*$. Numerical specification of hue and saturation has been standardized by the CIE, but is not useful for color specification or color image coding.

A color specification system needs to be able to represent any color with high precision. Since few colors are handled at a time, a specification system can be computationally complex. A system for color specification must be intimately related to the CIE system.

Figure 7.1 **Color systems** can be classified into four groups that are related by different kinds of transformations. Tristimulus systems and perceptually uniform systems are useful for image coding.

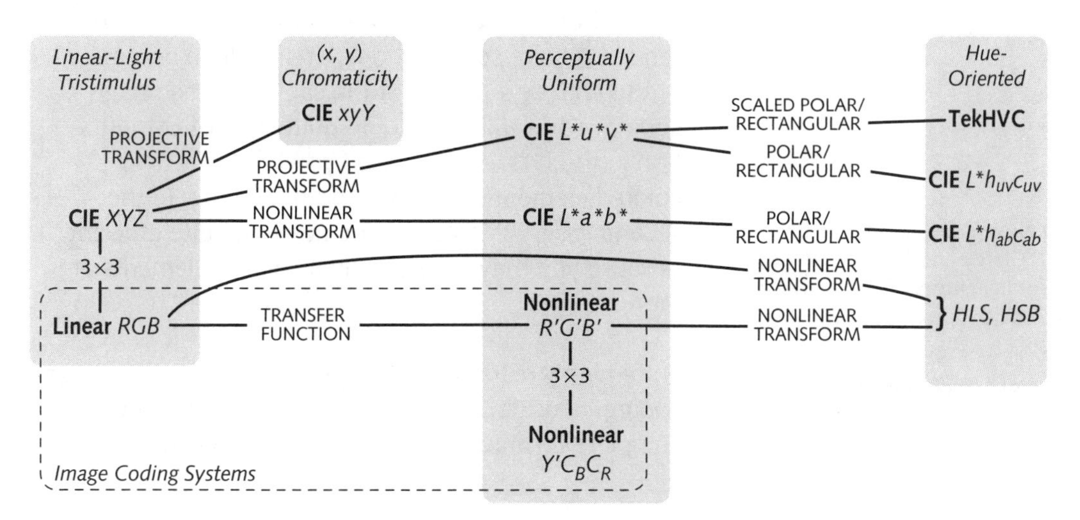

You can specify an ink color by specifying the proportions of standard (or secret) inks that can be mixed to make the color. That technique is not useful for image coding.

Color image coding

A color image is represented as an array of pixels, where each pixel contains numerical components that define a color. Three components are necessary and sufficient for this purpose, although in printing it is convenient to use a fourth (black) component.

In theory, the three numerical values for image coding could be provided by a color specification system. But a practical image coding system needs to be computationally efficient, cannot afford unlimited precision, need not be intimately related to the CIE system, and generally needs to cover only a reasonably wide range of colors and not all possible colors. So image coding uses different systems than color specification.

The systems useful for image coding are linear *RGB*, nonlinear *R'G'B'*, nonlinear *CMY*, nonlinear *CMYK*, and derivatives of nonlinear *R'G'B'*, such as $Y'C_BC_R$. Numerical values of hue and saturation are not useful in color image coding.

If you manufacture cars, you have to match the color of paint on the door with the color of paint on the fender. A color specification system will be necessary. But to convey a picture of the car, you need image coding. You can afford to do quite a bit of computation in the first case, because you have only two colored elements, the door and the fender. In the second case, the color coding must be quite efficient, because you may have a million colored elements or more.

LeRoy E. DeMarsh and Edward J. Giorgianni, "Color Science for Imaging Systems," *Physics Today*, September 1989, 44–52.

William F. Schreiber, *Fundamentals of Electronic Imaging Systems*, Second Edition. Berlin: Springer-Verlag, 1991.

For a highly readable short introduction to color image coding, consult DeMarsh and Giorgianni. For a terse, complete technical treatment, read Schreiber.

Definitions

Intensity is the rate at which radiant energy is transferred per unit area. In image science, radiant power is measured over some interval of the electromagnetic spectrum, and we're interested in power radiating from or incident on a surface. Intensity is what I call a *linear-light* measure, expressed in units such as watts per square meter.

Brightness is defined by the CIE as *the attribute of a visual sensation according to which an area appears to exhibit more or less light*. Because brightness perception is very complex, the CIE defined a more tractable quantity *luminance,* which is radiant power weighted by a spectral sensitivity function that is characteristic of vision.

Hue is the attribute of a color perception denoted by blue, green, yellow, red, and so on. Roughly speaking, if the dominant wavelength of a spectral power distribution shifts, the hue of the associated color will shift.

Saturation – or *purity* – is the degree of colorfulness, from neutral gray through pastel to saturated colors. Roughly speaking, the more an SPD is concentrated at one wavelength, the more saturated will be the associated color. You can desaturate a color by adding light that contains power at all wavelengths.

Spectral power distribution (SPD) and tristimulus

Physical power (or *radiance*) is expressed in a *spectral power distribution* (SPD), often in 31 components, each representing a 10 nm band. The SPD of the CIE standard daylight illuminant CIE D_{65} is graphed at the upper left of Figure 7.2 overleaf.

One way to describe a color is to directly reproduce its spectral power distribution. In Figure 7.2 overleaf, 31 components are sketched. This method is reasonable to describe a single color or a few colors, but using 31 components for each pixel is an impractical way to code an image. Due to the trichromatic nature of

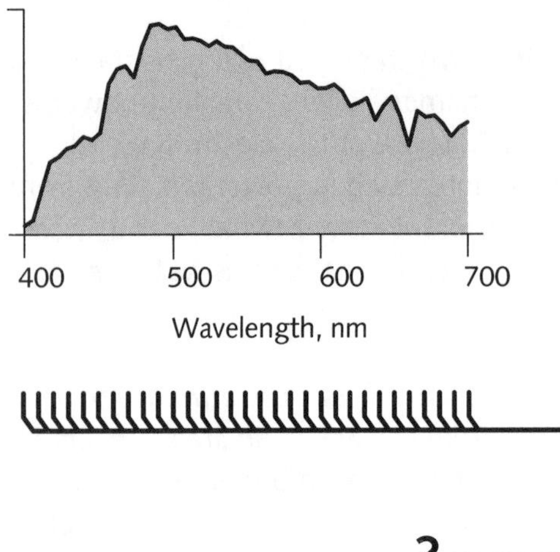

Figure 7.2 **Spectral and tristimulus color reproduction.** A color can be described as a spectral power distribution (SPD), perhaps in 31 components representing power in 10 nm intervals over the range 400 nm to 700 nm. But if appropriate spectral weighting functions are used, three values are sufficient. The SPD shown here is the D_{65} daylight illuminant standardized by the CIE.

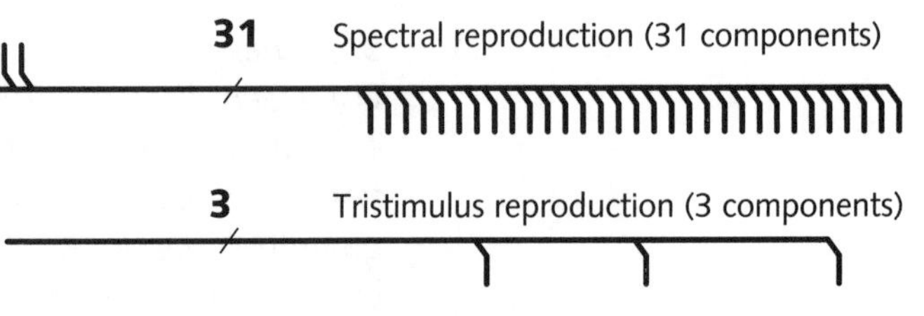

vision, a suitable spectral weighting functions can be used to describe a color using just three components.

The relationship between SPDs and perceived color is the concern of the science of *colorimetry*. In 1931, the *Commission Internationale de L'Éclairage* (CIE) adopted functions for a hypothetical *Standard Observer*, and those functions specify how an SPD can be transformed into a triple that specifies a color.

The CIE system applies directly to self-luminous sources and displays. However, the colors produced by reflective systems, such as photography, printing, or paint, depend not only on the colorants, but also on the SPD of the illumination. If your application has a strong dependence upon the spectrum of the illuminant, you may have to resort to spectral matching.

Scanner spectral constraints

The relationship between spectral distributions and the three components of a color value is conventionally explained starting from the famous color matching experiment. I will instead explain the relationship by illustrating the practical concerns of engineering the spectral filters required by a color scanner or camera, using Figure 7.3 opposite.

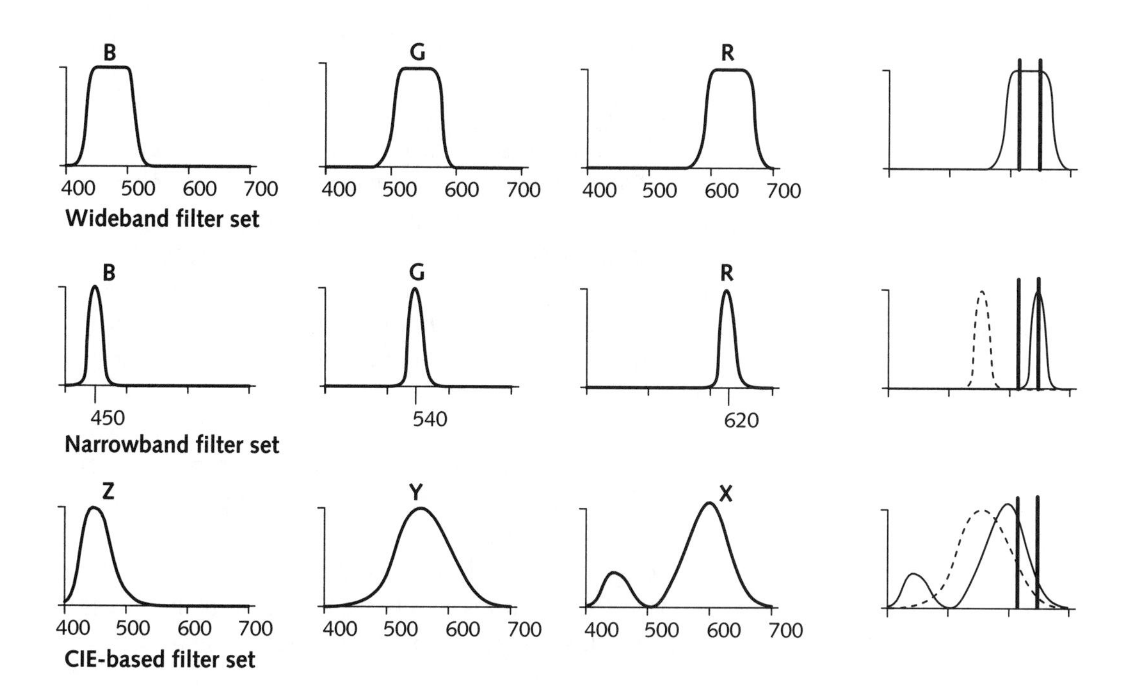

Figure 7.3 **Scanner spectral constraints.** This diagram shows the spectral constraints associated with scanners and cameras. The **wideband filter set** of the top row shows the spectral sensitivity of filters having uniform response across the shortwave, mediumwave, and longwave regions of the spectrum. With this approach, two monochromatic sources seen by the eye to have different colors – in this case, saturated orange and a saturated red – cannot be distinguished by the filter set. **The narrowband filter set** in the middle row solves that problem, but creates another: Many monochromatic sources "fall between" the filters, and are seen by the scanner as black. To see color as the eye does, the three filter responses must be closely related to the color response of the eye. **The CIE-based filter set** in the bottom row shows the *color matching functions* (CMFs) of the CIE Standard Observer.

The top row shows the spectral sensitivity of three wideband optical filters having uniform response across each of the longwave, mediumwave, and shortwave regions of the spectrum. In most filter applications, whether for electrical signals or for optical power, it is desirable to have a response that is as uniform as possible across the passband, to have transition zones as narrow as possible, and to have maximum possible attenuation in the stopbands. Textbooks on filter design concentrate on filters optimized in that style. The top row of the illustration shows two monochromatic sources, which appear saturated orange and red, analyzed by "textbook" bandpass filters. These two different wavelength distributions, which are seen as different colors, report the identical *RGB* triple [1, 0, 0].

At first sight it may seem that the problem with the wideband filters is insufficient wavelength discrimination. The middle row of the example shows three narrowband filters. This set solves one problem, but creates another: Many monochromatic sources "fall between" the filters, and are seen by the scanner as black. In the example, the orange source reports an *RGB* triple of [0, 0, 0], identical to the result of scanning black.

Although this example is contrived, the problem is not. Ultimately, the test of whether a camera or scanner is successful is whether it reports distinct *RGB* triples if and only if human vision sees different colors. To see color as the eye does, the three filter sensitivity curves must be closely related to the color response of the eye.

Colorimetry, Second Edition, Publication CIE N° 15.2. Vienna, Austria: Central Bureau of the Commission Internationale de L'Éclairage, 1986.

A famous experiment, the *color matching experiment*, was devised during the 1920s to characterize the relationship between physical spectra and perceived color. The experiment measures mixtures of different spectral distributions that are required for human observers to match colors. Exploiting this indirect method, the CIE in 1931 standardized a set of spectral weighting functions that models the perception of color. These curves, defined numerically, are referred to as the $\bar{x}$, $\bar{y}$, and $\bar{z}$ *color matching functions* (*CMFs*) for the CIE Standard Observer. They are illustrated in the bottom row of Figure 7.3, and are graphed at a larger scale in Figure 7.4 opposite.

$\bar{x}$, $\bar{y}$, and $\bar{z}$ are pronounced "*x-bar, y-bar, z-bar.*"

CIE XYZ tristimulus

The luminance *Y* of a source is obtained by integrating its SPD weighted by the $\bar{y}$ color matching function, as explained in *Luminance and lightness*, on page 81.

When luminance is augmented with two other components *X* and *Z* similarly computed using the $\bar{x}$ and $\bar{z}$ color matching functions, the resulting (*X*, *Y*, *Z*)

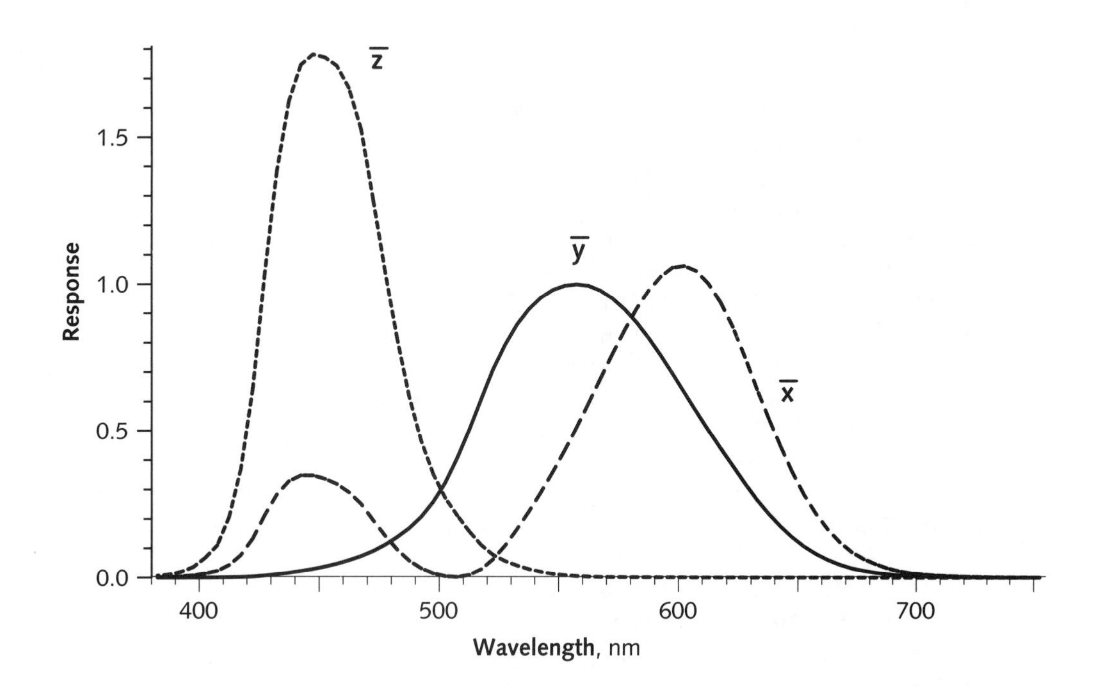

Figure 7.4 **CIE color matching functions.** A camera must have these spectral response curves, or linear combinations of them, in order to capture all colors. However, practical considerations make this difficult. These analysis functions are not comparable to spectral power distributions.

components – pronounced "*big-X, big-Y, big-Z*" or "*cap-X, cap-Y, cap-Z*" – are known as *XYZ tristimulus* values. These are linear-light values that embed the spectral properties of human color vision.

Tristimulus values are computed from continuous SPDs by integrating the SPD using the $\bar{x}$, $\bar{y}$, and $\bar{z}$ color matching functions. In discrete form, tristimulus values can be computed by a matrix multiplication, as illustrated in Figure 7.5 overleaf.

Grassmann's Third Law:

Sources of the same color produce identical effects in a mixture regardless of their spectral composition.

Human color vision follows a principle of superposition now known as Grassmann's Third Law: The tristimulus values computed from the sum of a set of SPDs is identical to the sum of the tristimulus values of each SPD. Due to this linearity of additive color mixture, any set of three components that is a nontrivial linear combination of X, Y, and Z is also a set of tristimulus values.

$$
\begin{bmatrix} X \\ Y \\ Z \end{bmatrix} =
\begin{bmatrix}
0.0143 & 0.0004 & 0.0679 \\
0.0435 & 0.0012 & 0.2074 \\
0.1344 & 0.0040 & 0.6456 \\
0.2839 & 0.0116 & 1.3856 \\
0.3483 & 0.0230 & 1.7471 \\
0.3362 & 0.0380 & 1.7721 \\
0.2908 & 0.0600 & 1.6692 \\
0.1954 & 0.0910 & 1.2876 \\
0.0956 & 0.1390 & 0.8130 \\
0.0320 & 0.2080 & 0.4652 \\
0.0049 & 0.3230 & 0.2720 \\
0.0093 & 0.5030 & 0.1582 \\
0.0633 & 0.7100 & 0.0782 \\
0.1655 & 0.8620 & 0.0422 \\
0.2904 & 0.9540 & 0.0203 \\
0.4334 & 0.9950 & 0.0087 \\
0.5945 & 0.9950 & 0.0039 \\
0.7621 & 0.9520 & 0.0021 \\
0.9163 & 0.8700 & 0.0017 \\
1.0263 & 0.7570 & 0.0011 \\
1.0622 & 0.6310 & 0.0008 \\
1.0026 & 0.5030 & 0.0003 \\
0.8544 & 0.3810 & 0.0002 \\
0.6424 & 0.2650 & 0.0000 \\
0.4479 & 0.1750 & 0.0000 \\
0.2835 & 0.1070 & 0.0000 \\
0.1649 & 0.0610 & 0.0000 \\
0.0874 & 0.0320 & 0.0000 \\
0.0468 & 0.0170 & 0.0000 \\
0.0227 & 0.0082 & 0.0000 \\
0.0114 & 0.0041 & 0.0000
\end{bmatrix}^{\mathsf{T}}
\bullet
\begin{bmatrix}
82.75 \\ 91.49 \\ 93.43 \\ 86.68 \\ 104.86 \\ 117.01 \\ 117.81 \\ 114.86 \\ 115.92 \\ 108.81 \\ 109.35 \\ 107.80 \\ 104.79 \\ 107.69 \\ 104.41 \\ 104.05 \\ 100.00 \\ 96.33 \\ 95.79 \\ 88.69 \\ 90.01 \\ 89.60 \\ 87.70 \\ 83.29 \\ 83.70 \\ 80.03 \\ 80.21 \\ 82.28 \\ 78.28 \\ 69.72 \\ 71.61
\end{bmatrix}
\begin{matrix}
\text{400 nm} \\ \\ \\ \\ \\ \text{450 nm} \\ \\ \\ \\ \\ \text{500 nm} \\ \\ \\ \\ \\ \text{550 nm} \\ \\ \\ \\ \\ \text{600 nm} \\ \\ \\ \\ \\ \text{650 nm} \\ \\ \\ \\ \\ \text{700 nm}
\end{matrix}
$$

Figure 7.5 **Calculation of tristimulus values by matrix multiplication.** The 31-element column vector at the right is a discrete version of CIE Illuminant D_{65}, at 10 nm intervals. The 31×3 matrix is a discrete version of the CIE $\bar{x}$, $\bar{y}$, and $\bar{z}$ color matching functions. Performing the matrix multiplication produces a set of three *XYZ* tristimulus components.

The CIE system is based on the description of color as a luminance component Y, as described above, and two additional components X and Z. The spectral weighting curves of X and Z have been standardized by the CIE based on statistics from experiments involving human observers. *XYZ tristimulus values* can describe any color. (*RGB* tristimulus values will be described on page 133.)

The values of the *XYZ* components are proportional to physical power, but their spectral composition corresponds to the color matching characteristics of human vision. To describe a color, it is not necessary to specify its spectrum: it suffices to have three tristimulus components, computed according to the principles of the CIE.

CIE (x, y) chromaticity

It is convenient, for both conceptual understanding and computation, to have a representation of "pure" color in the absence of brightness. The CIE standardized a procedure for normalizing *XYZ* tristimulus values to obtain two *chromaticity* values x and y (pronounced "*little-x, little-y*"). The relationships are computed by this projective transformation:

Eq 7.1

$$x = \frac{X}{X+Y+Z} \qquad y = \frac{Y}{X+Y+Z}$$

The chromaticity coordinates x and y are abstract entities that have no direct physical interpretation. A color plots as a point in an (x, y) *chromaticity diagram*, illustrated in Figure 7.6 overleaf. A third chromaticity, z, is defined, but is redundant since $x + y + z = 1$.

A color can be specified by its chromaticity and luminance, in the form of an *xyY* triple. To recover X and Z tristimulus values from (x, y) chromaticities and luminance, use the inverse of Equation 7.1:

Eq 7.2

$$X = \frac{x}{y} Y \qquad Z = \frac{1 - x - y}{y} Y$$

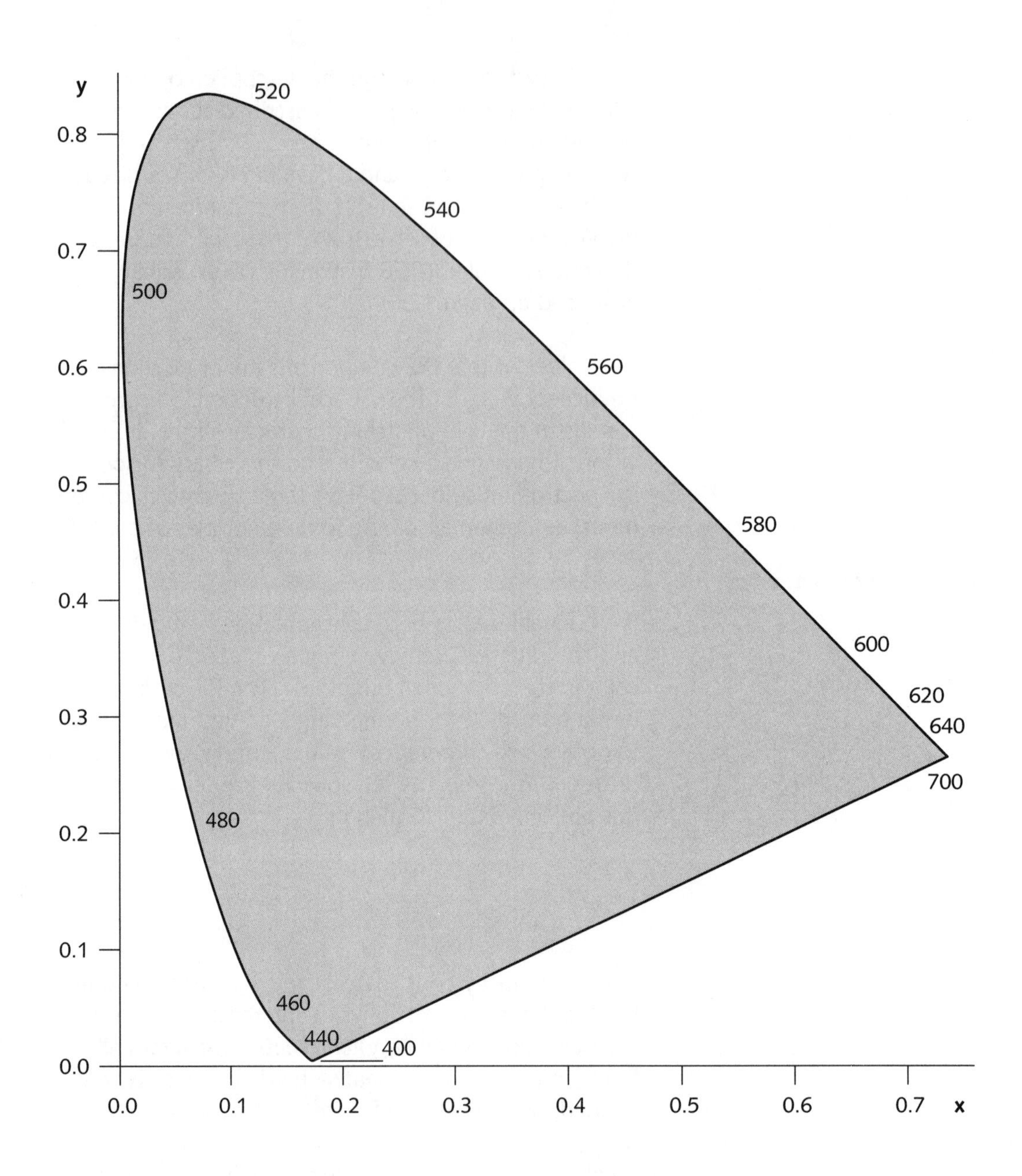

Figure 7.6 **CIE (x, y) chromaticity diagram.** The *spectral locus* is an inverted U-shaped path swept by a monochromatic source as it is tuned from 400 nm to 700 nm. The set of all colors is closed by the *line of purples*, which traces SPDs that combine longwave and shortwave power but have no mediumwave contribution. There is no unique definition of white, but it lies near the center of the chart. All colors lie within the U-shaped region: points outside this region are not associated with colors.

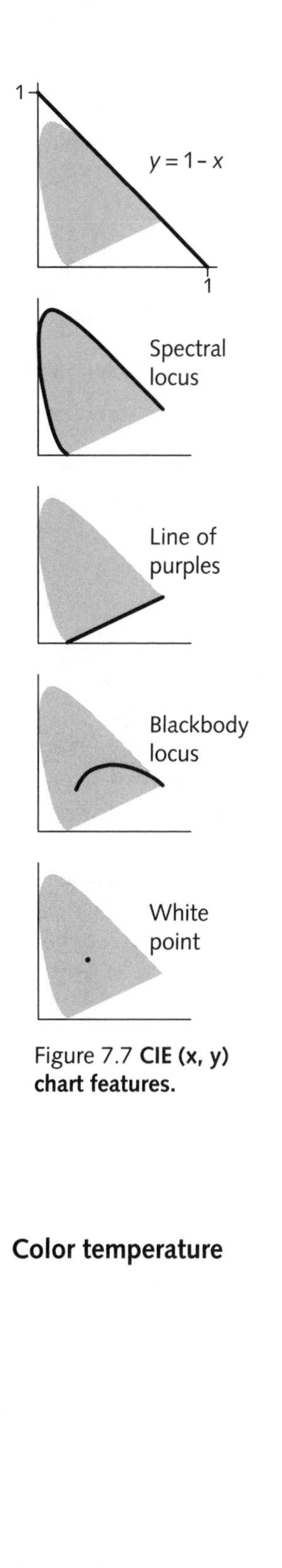

Figure 7.7 **CIE (x, y) chart features.**

When a narrowband (monochromatic) SPD comprising power at just one wavelength is swept across the range 400 nm to 700 nm, it traces a shark-fin-shaped *spectral locus* in (x, y) coordinates.

The sensation of purple cannot be produced by a single wavelength; it requires a mixture of shortwave and longwave light. The *line of purples* on a chromaticity diagram joins the chromaticity of extreme blue, comprising only shortwave power, to the chromaticity of extreme red, comprising only longwave power

Any all-positive (*physical*, or *realizable*) SPD plots as a single point in the chromaticity diagram, within or on the boundary of the spectral locus and the line of purples. All colors are within this region: points outside this region are not associated with colors.

In the section *White*, on page 129, I will describe the SPDs associated with white. There is no unique physical or perceptual definition of white. An SPD that is considered white will have CIE (x, y) coordinates roughly in the central area of the chromaticity diagram, in the region of $(\frac{1}{3}, \frac{1}{3})$. Many important sources of illumination have chromaticity coordinates that lie on the *blackbody locus*; the SPDs of blackbody radiators will be discussed in the next section.

In the projective transformation that forms x and y, any additive mixture (linear combination) of two SPDs – or two tristimulus values – plots on a straight line in the (x, y) plane. But distances are not preserved, so chromaticity values do not combine linearly.

Color temperature

Max Planck determined that the SPD radiated from a hot object – a *blackbody radiator* – is a function of the temperature to which the object is heated. Many sources of illumination have, at their core, a heated object, so it is often useful to characterize an illuminant by specifying the absolute temperature (in units of kelvin, K) of a blackbody radiator that appears to have

Figure 7.8 **SPDs of blackbody radiators** at several temperatures are graphed here. As the temperature increases, the absolute power increases and the peak of the spectral distribution shifts toward shorter wavelengths.

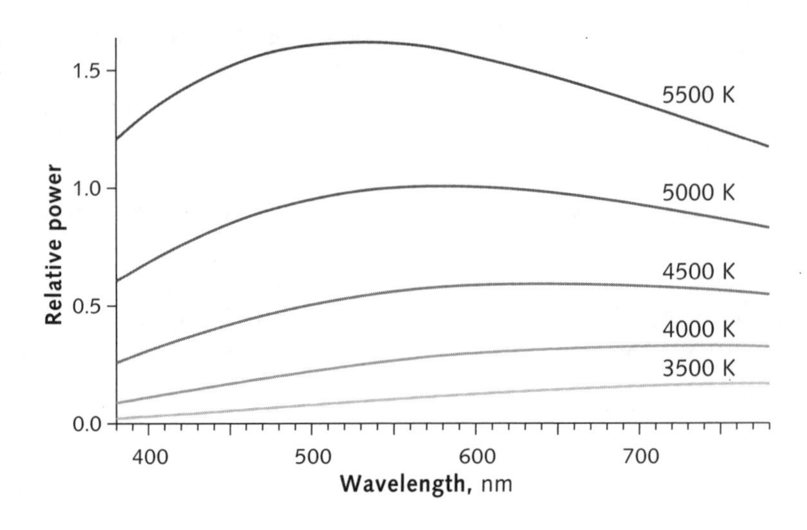

the same hue. Figure 7.8 above shows the SPDs of blackbody radiators at several temperatures. As temperature increases, the absolute power increases and the spectral peak shifts toward shorter wavelengths. If the power of blackbody radiators is normalized at some arbitrary wavelength, dramatic differences in spectral character are evident, as illustrated in Figure 7.9 below.

Figure 7.9 **SPDs of blackbody radiators, normalized** to equal power at 560 nm, are graphed here. The dramatically different spectral character of different blackbody radiators is evident.

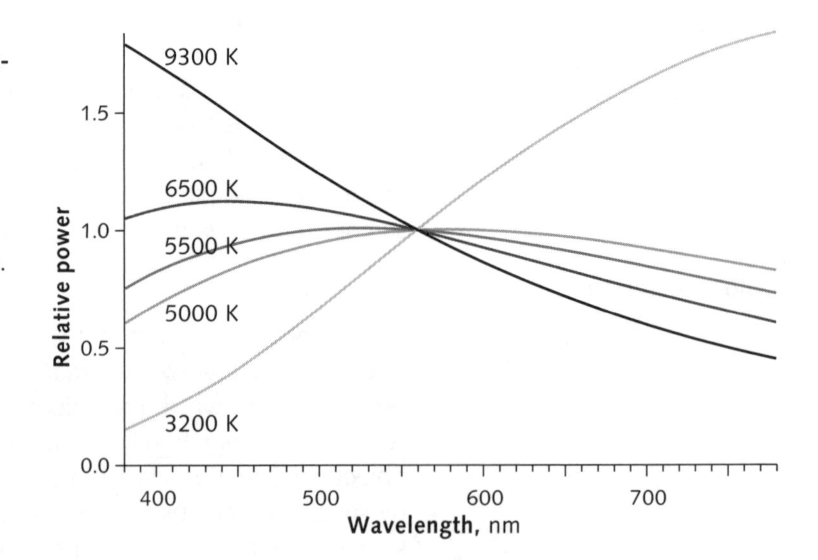

White

In additive image reproduction, to be detailed on page 132, the *white point* is the color reproduced by equal red, green, and blue components. The color of white is a function of the ratio (or *balance*) of power among the primaries. In subtractive reproduction, to be detailed on page 151, white is the SPD of the illumination, multiplied by the SPD of the media. There is no unique physical or perceptual definition of white, so to achieve accurate color interchange you must specify the characteristics of your white.

It is convenient for purposes of calculation to define white as an SPD whose power is uniform throughout the visible spectrum. This white reference is known as the *equal-energy illuminant*, or *CIE Illuminant E*.

A more realistic reference that approximates daylight has been specified numerically by the CIE as Illuminant D_{65}. You should use this unless you have a good reason to use something else. The print industry commonly uses D_{50} and photography commonly uses D_{55}. These represent compromises between the conditions of indoor (tungsten) and daylight viewing. Figure 7.10 overleaf shows the SPDs of several standard illuminants.

A white reference of 9300 K contains far too much blue to be useful for high quality image reproduction. See page 150.

An illuminant can be specified informally by its color temperature. Sometimes the white reference of a computer monitor is specified as a color temperature, often 9300 K, along with an indication of the displacement of the white point from the blackbody locus. A more accurate and direct specification is provided by the chromaticity coordinates of the SPD of the source.

Human vision adapts to the viewing environment. An image viewed in isolation – such as a slide projected in a dark room – creates its own white reference, and a viewer will be quite tolerant variation in the white point. But if the same image is viewed in the presence of an external white reference or a second image, then differences in white point can be objectionable.

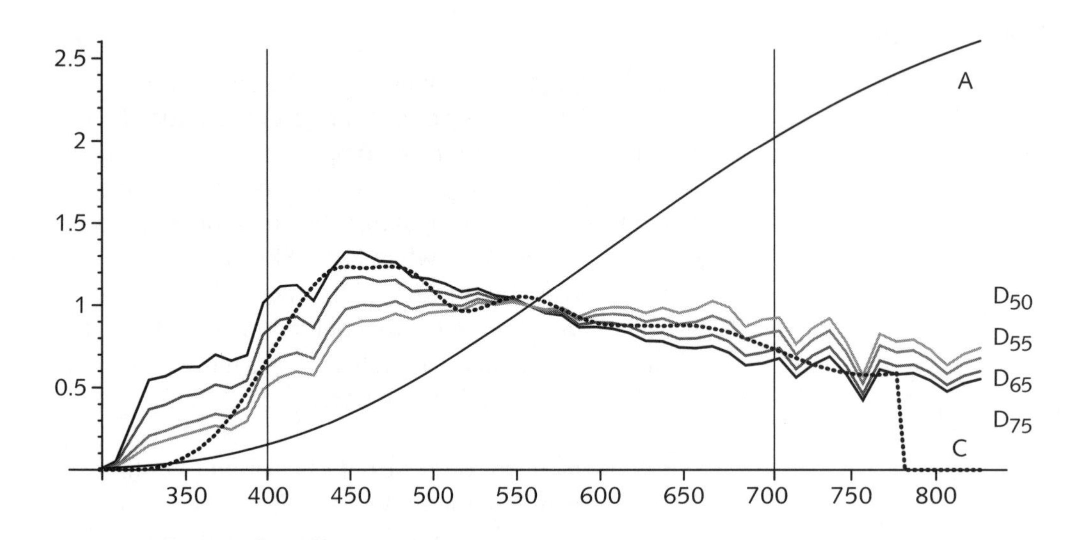

Figure 7.10 **CIE illuminants** are graphed here. Illuminant A is an obsolete standard representative of tungsten illumination; its SPD resembles the blackbody radiator at 3200 K shown in Figure 7.9, on page 128. Illuminant C was an early standard for daylight; it too is obsolete. The family of D illuminants represents daylight at several color temperatures.

Complete adaptation seems to be confined to color temperatures from 5000 K to 5500 K. For most people, under most viewing conditions, D_{65} has a little hint of blue. Tungsten illumination, at about 3200 K, almost always appears somewhat yellow.

Perceptually uniform color spaces

As outlined in *Perceptual uniformity*, on page 17, a system is perceptually uniform if a small perturbation to a component value is approximately equally perceptible across the range of that value.

XYZ and *RGB* tristimulus values, and *xyY* chromaticity and luminance, are far from exhibiting perceptual uniformity. Finding a transformation of *XYZ* into a reasonably perceptually uniform space consumed a decade or more at the CIE, and in the end no single system could be agreed upon. So the CIE standardized two systems, $L^*u^*v^*$ and $L^*a^*b^*$, sometimes written CIELUV and CIELAB. (The *u* and *v* or CIE are unrelated to *U* and *V* of video.) Both $L^*u^*v^*$ and $L^*a^*b^*$ improve the 80:1 or so perceptual nonuniformity of

XYZ to about 6:1. Both demand too much computation to accommodate real-time display, although both have been successfully applied to image coding for printing.

As described on page 88, the perceptual response to luminance is approximated as *Lightness*, denoted L^*, defined by the CIE as:

Eq 7.3

$$L^* = \begin{cases} 903.3\,\frac{Y}{Y_n}, & \frac{Y}{Y_n} \le 0.008856 \\ 116\left(\frac{Y}{Y_n}\right)^{\frac{1}{3}} - 16, & 0.008856 < \frac{Y}{Y_n} \end{cases}$$

Y_n is the luminance of the white reference. L^* has a range of 0 to 100, and a difference of unity between two L^* values – ΔL^*, pronounced *delta L-star* – is roughly the threshold of discrimination. Video systems approximate this lightness response, by using $R'G'B'$ signals that are each subject to a 0.45 power function comparable to the $1/3$ power function defined by L^*.

Computation of CIE $L^*u^*v^*$ involves intermediate u' and v' quantities, where the prime denotes the successor to the obsolete 1960 CIE u and v system:

Eq 7.4

$$u' = \frac{4\,X}{X + 15\,Y + 3\,Z}, \qquad v' = \frac{9\,Y}{X + 15\,Y + 3\,Z}$$

First compute u_n' and v_n' for your reference white X_n, Y_n, and Z_n. Then compute u' and v' – and L^*, as discussed earlier – for your colors. Finally, compute:

Eq 7.5

$$u^* = 13L^*\left(u' - u_n'\right), \qquad v^* = 13L^*\left(v' - v_n'\right)$$

a^* and b^* are computed as follows, providing that X/X_n, Y/Y_n, and Z/Z_n are all greater than 0.008856:

Eq 7.6

$$a^* = 500\left[\left(\frac{X}{X_n}\right)^{\frac{1}{3}} - \left(\frac{Y}{Y_n}\right)^{\frac{1}{3}}\right], \qquad b^* = 200\left[\left(\frac{Y}{Y_n}\right)^{\frac{1}{3}} - \left(\frac{Z}{Z_n}\right)^{\frac{1}{3}}\right]$$

Values of X/X_n, Y/Y_n, and Z/Z_n as small as 0.008856 are not normally encountered in image coding, but a linear segment of the transfer function is defined by the CIE to accommodate values below this threshold. The complexity of these calculations makes CIE $L^*u^*v^*$ and CIE $L^*a^*b^*$ unsuitable for image coding. Although not specifically optimized for this purpose, the nonlinear $R'G'B'$ coding used in video is quite perceptually uniform, and has the advantage of being fast enough for interactive applications.

Additive mixture, RGB

Previous sections explained how a physical SPD can be analyzed into three components that represent color. This section explains how three components can be used to reproduce color.

The simplest way to reproduce a range of colors is to mix the beams from lights of three different colors. The widest range of colors will be reproduced with red, green, and blue lights. Figure 7.11 opposite illustrates additive reproduction. In physical terms, the spectra from each of the lights add together wavelength by wavelength to form the spectrum of the mixture.

As a consequence of the principle of superposition, the color of an additive mixture is a strict function of the colors of the primaries and the fraction of each primary that is mixed. The SPDs of the primaries need not be known, just their tristimulus values, or their luminance values and chromaticity coordinates.

Additive reproduction is employed directly in a video projector, where the spectra from a red beam, a green beam, and a blue beam are physically summed at the surface of the projection screen. Additive reproduction is also employed in a direct-view color CRT, but through slightly indirect means. The screen of a CRT comprises small dots that produce red, green, and blue light. When the screen is viewed from a sufficient distance, the spectra of these dots add at the retina of the observer.

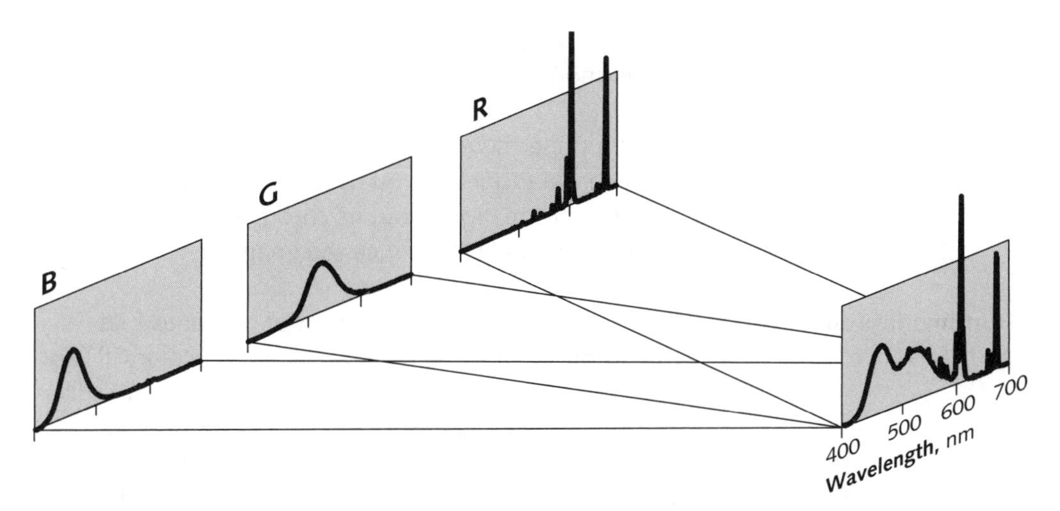

Figure 7.11 **Additive mixture.** This diagram illustrates the physical process underlying additive color mixture, as is used in color television. Each colorant has an independent, direct path to the image. The spectral power of the image is, at each wavelength, the sum of the spectra of the colorants. The colors of the mixtures are completely determined by the colors of the primaries; analysis and prediction of mixtures is reasonably simple. The SPDs shown here are those of a Sony Trinitron monitor.

Characterization of RGB primaries

Additive reproduction is based on physical devices that produce all-positive SPDs for each primary. Physically and mathematically, the spectra add. The largest range of colors will be produced with primaries that appear red, green, and blue. Color vision obeys the principle of superposition, so the color produced by any additive mixture of three primary spectra can be predicted by adding the corresponding fractions of the *XYZ* components of the primaries: The colors that can be mixed from a particular set of *RGB* primaries are completely determined by the colors of the primaries by themselves. Subtractive reproduction is much more complicated, as will be explained on page 151: The colors of subtractive mixtures are determined not only by the primaries, but also by the colors of their combinations.

An additive *RGB* system is specified by the chromaticities of its primaries and its white point. The extent – or *gamut* – of the colors that can be mixed from a given set of *RGB* primaries is given in the (x, y) chromaticity diagram by a triangle whose vertices are the chromaticities of the primaries.

Figure 7.12 opposite plots the primaries of several contemporary video standards to be described.

In computing there are no standard primaries or white point chromaticities. If you have an *RGB* image but have no information about its chromaticities, you cannot accurately reproduce the image.

NTSC primaries (obsolete)

The NTSC in 1953 specified a set of primaries that were representative of phosphors used in color CRTs of that era. Those primaries and white reference are still documented in ITU-R Report 624. But phosphors changed over the years, primarily in response to market pressures for brighter receivers, and by the time of the first videotape recorder the primaries actually in use were quite different from those "on the books." So although you may see the NTSC primary chromaticities documented – even in standards for image exchange – they are of no practical use today. I include them in Table 7.1, for reference:

Table 7.1 **NTSC primaries (obsolete)**

	Red	*Green*	*Blue*	*White CIE Ill. C*
x	0.67	0.21	0.14	0.310
y	0.33	0.71	0.08	0.316
z	0.	0.08	0.78	0.374

EBU primaries

EBU Tech. 3213, *EBU standard for chromaticity tolerances for studio monitors.* Geneva: European Broadcasting Union, 1975 (re-issued 1981).

By the time 625/50 color video was standardized by the European Broadcasting Union (EBU), phosphor technology had improved considerably from the days of the NTSC. These primaries in Table 7.2 below are standardized by EBU Tech. 3213. They are in use today for 625/50 systems, and they are very close to the Rec. 709 primaries to be described in a moment:

Table 7.2 **EBU primaries**

	Red	*Green*	*Blue*	*White, D_{65}*
x	0.640	0.290	0.150	0.3127
y	0.330	0.600	0.060	0.3290
z	0.030	0.110	0.790	0.3582

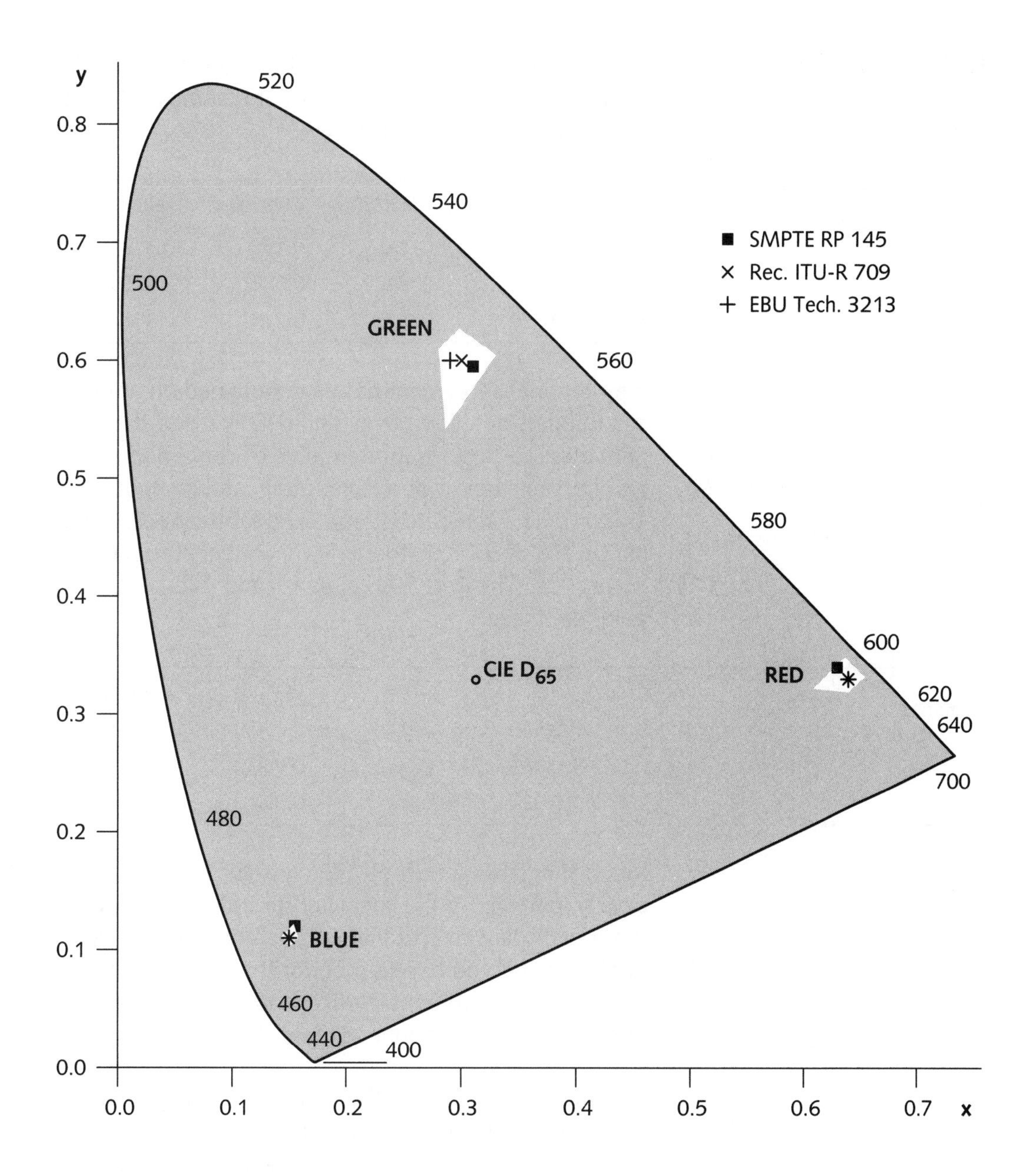

Figure 7.12 **RGB primaries of video standards** are plotted on the CIE (*x*, *y*) chromaticity diagram. The colors that can be represented in positive *RGB* values lie within the triangle formed by the primaries. The Rec. 709 standard specifies no tolerance. SMPTE tolerances are specified as ±0.005 in *x* and *y*. EBU tolerances are shown as white quadrilaterals; they are specified in *u′*, *v′* coordinates related to the color discrimination of vision. The EBU tolerance boundaries are not parallel to the (*x*, *y*) axes.

SMPTE RP 145 primaries

SMPTE RP 145-1994, *SMPTE C Color Monitor Colorimetry.*

For 525/59.94 systems, and for 1125/60 1920×1035 HDTV according to SMPTE 240M, it is standard to use the primaries of SMPTE RP 145:

Table 7.3 **SMPTE RP 145 primaries**

	Red	*Green*	*Blue*	*White,* D_{65}
x	0.630	0.310	0.155	0.3127
y	0.340	0.595	0.070	0.3290
z	0.030	0.095	0.775	0.3582

Rec. 709 primaries

Recommendation ITU-R BT.709, *Basic Parameter Values for the HDTV Standard for the Studio and for International Programme Exchange*. Geneva: ITU, 1990.

International agreement has been obtained on primaries for high-definition television (HDTV), and these primaries are closely representative of contemporary monitors in studio video, computing, and computer graphics. The standard is formally denoted *Recommendation ITU-R BT.709* (formerly CCIR Rec. 709). I'll call it *Rec. 709*. The Rec. 709 primaries and its D_{65} white point are these:

Table 7.4 **Rec. 709 primaries**

	Red	*Green*	*Blue*	*White,* D_{65}
x	0.640	0.300	0.150	0.3127
y	0.330	0.600	0.060	0.3290
z	0.030	0.100	0.790	0.3582

Video standards specify *R'G'B'* systems that are closely matched to the characteristics of real monitors. Physical display devices involve tolerances and uncertainties, but if you have a monitor that conforms to Rec. 709 within some tolerance, you can consider the monitor to be device-independent.

The importance of Rec. 709 as an interchange standard in studio video, broadcast television, and high-definition television, and the firm perceptual basis of the standard, assures that its parameters will be used even by such devices as flat-panel displays that do not have the same physics as CRTs.

CMFs and SPDs

You might guess that you could implement a display whose primaries had spectral power distributions with the same shape as the CIE spectral analysis curves – the color matching functions for *XYZ*. You could make such a display, but when driven by *XYZ* tristimulus values, it would not properly reproduce color. There are display primaries that reproduce color accurately when driven by *XYZ* tristimuli, but the SPDs of those primaries do not have the same shape as the $\bar{x}$, $\bar{y}$, and $\bar{z}$ CMFs. To see why requires understanding a very subtle and important point about the CIE system as implemented in real devices.

To find a set of display primaries that reproduce color according to *XYZ* tristimulus values requires constructing three SPDs, which, when analyzed by the $\bar{x}$, $\bar{y}$, and $\bar{z}$ color matching functions, produce [1, 0, 0], [0, 1, 0], and [0, 0, 1], respectively. The $\bar{x}$, $\bar{y}$, and $\bar{z}$ CMFs are positive across the entire spectrum. Producing [0, 1, 0] will require positive contribution from some wavelengths in the required primary SPDs, and that we could arrange, but there is no wavelength that contributes to *Y* that does not also contribute positively to *X* or *Z*.

The only mathematical solution to this dilemma is to force the *X* and *Z* contributions to zero by making the corresponding SPDs have negative power at certain wavelengths. Although this is not a problem for mathematics, or even for signal processing, an SPD with a negative portion is not physically realizable. So we cannot build a real display that responds directly to *XYZ*. But as you will see, the concept of negative SPDs – and *nonphysical SPDs* or *nonrealizable primaries* – is very useful in theory and in practice.

If you want to understand the mathematical details of color transforms, described on the following ten pages, you should be familiar with linear (matrix) algebra. If you are unfamiliar with linear algebra, see Howard Anton and Chris Rorres, *Elementary Linear Algebra*, Seventh Edition. New York: John Wiley & Sons, 1994.

There are many ways to choose nonphysical primary SPDs that correspond to the $\bar{x}$, $\bar{y}$, and $\bar{z}$ color matching functions. One way is to arbitrarily choose three display primaries whose power is concentrated at the same three discrete wavelengths. Consider three display SPDs that are combinations of power at 600 nm, 550 nm, and 470 nm. Sample the $\bar{x}$, $\bar{y}$, and $\bar{z}$ functions of the matrix given earlier in *Calculation of tristimulus values by matrix multiplication*, on page 124, at those three wavelengths:

	Red, 600 nm	*Green, 550 nm*	*Blue, 470 nm*
X	1.0622	0.4334	0.1954
Y	0.6310	0.9950	0.0910
Z	0.0008	0.0087	1.2876

These coefficients can be expressed as a matrix, where the column vectors give the *XYZ* tristimulus values corresponding to pure red, green, and blue at the display – [1, 0, 0], [0, 1, 0], and [0, 0, 1]. It is conventional to apply a scale factor in such a matrix to cause the middle row to sum to unity, since we require only relative matches, not absolute:

Eq 7.7

$$\begin{bmatrix} X \\ Y \\ Z \end{bmatrix} = \begin{bmatrix} 0.618637 & 0.252417 & 0.113803 \\ 0.367501 & 0.579499 & 0.052999 \\ 0.000466 & 0.005067 & 0.749913 \end{bmatrix} \bullet \begin{bmatrix} R_{600\,nm} \\ G_{550\,nm} \\ B_{470\,nm} \end{bmatrix}$$

That matrix gives the transformation from *RGB* to *XYZ*. We are interested in the inverse transform, from *XYZ* to *RGB*, so invert the matrix:

Eq 7.8

$$\begin{bmatrix} R_{600\,nm} \\ G_{550\,nm} \\ B_{470\,nm} \end{bmatrix} = \begin{bmatrix} 2.179151 & -0.946884 & -0.263777 \\ -1.382685 & 2.327499 & 0.045336 \\ 0.007989 & -0.015138 & 1.333346 \end{bmatrix} \bullet \begin{bmatrix} X \\ Y \\ Z \end{bmatrix}$$

The column vectors of this matrix give, for each primary, the weights of each of the three discrete wavelengths that are required to display unit *XYZ* tristimulus values. The color matching functions for CIE *XYZ* are shown in Figure 7.13, *CMFs for CIE XYZ primaries*, on page 140. Opposite those functions, in Figure 7.14, is The corresponding set of primary SPDs. As expected, the display primaries have some negative spectral components: The primary SPDs are nonphysical. Any set of primaries that reproduces color from *XYZ* tristimulus values is necessarily *supersaturated*, more saturated than any realizable SPD could be.

To determine a set of physical SPDs that will reproduce color when driven from *XYZ*, consider the problem in the other direction: Given a set of physically realizable display primaries, what CMFs are suitable to directly reproduce color using mixtures of these primaries? In this case the matrix that relates *RGB* components to CIE *XYZ* tristimulus values is all-positive, but the CMFs required for analysis of the scene have negative portions: The analysis filters are nonrealizable.

Figure 7.16 shows a set of primary SPDs conformant to SMPTE 240M, similar to Rec. 709. Many different SPDs can produce an exact match to these chromaticities. The set shown is from a Sony Trinitron monitor. Figure 7.15 shows the corresponding color matching functions. As expected, the CMFs have negative excursions and are not directly realizable.

We conclude that we can use physically realizable analysis CMFs, as in the first example, where *XYZ* components are displayed directly. But this requires nonphysical display primary SPDs. Or we can use physical display primary SPDs, but this requires nonphysical analysis CMFs. As a consequence of the way color vision works, there is no set of all-positive display primary SPDs that corresponds to an all-positive set of analysis functions.

The escape from this conundrum is not to use camera signals to directly drive the display, but to impose a 3×3 matrix multiplication in the processing of the camera signals. Consider these display primaries: monochromatic red at 600 nm, monochromatic green at 550 nm, and monochromatic blue at 470 nm. The 3×3 matrix of Equation 7.8 can be used to process *XYZ* values into components suitable to drive that display.

Figure 7.17 shows the set of spectral sensitivity functions implemented by the beam splitter and filter (*prism*) assembly of an actual video camera. The functions are positive everywhere across the spectrum, so the filters are physically realizable. However, rather

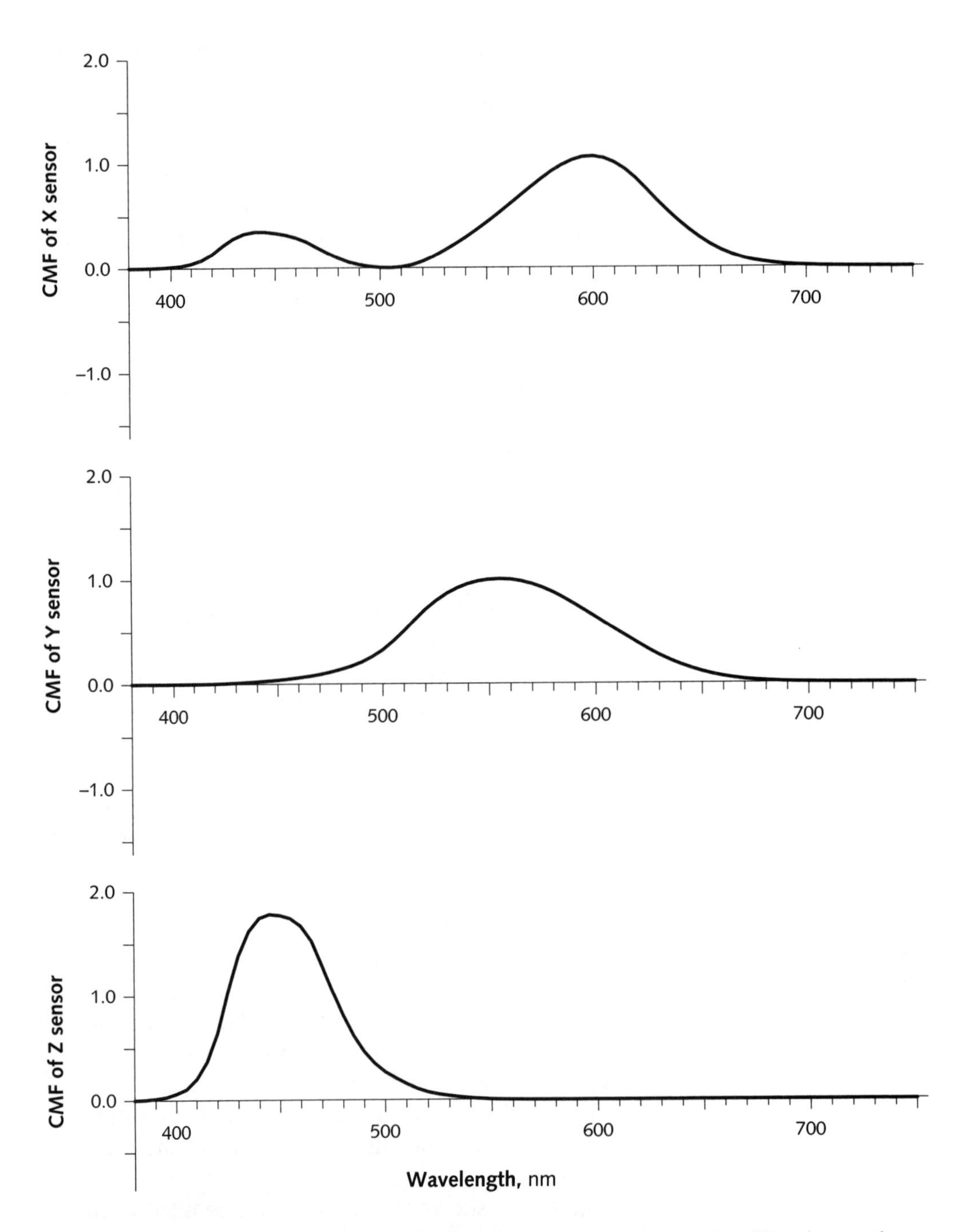

Figure 7.13 **CMFs for CIE XYZ primaries.** To acquire all colors in a scene requires filters having these spectral sensitivities. The functions are all-positive, and therefore can be realized in practice. However, these functions are seldom used in actual cameras or scanners, for various engineering reasons.

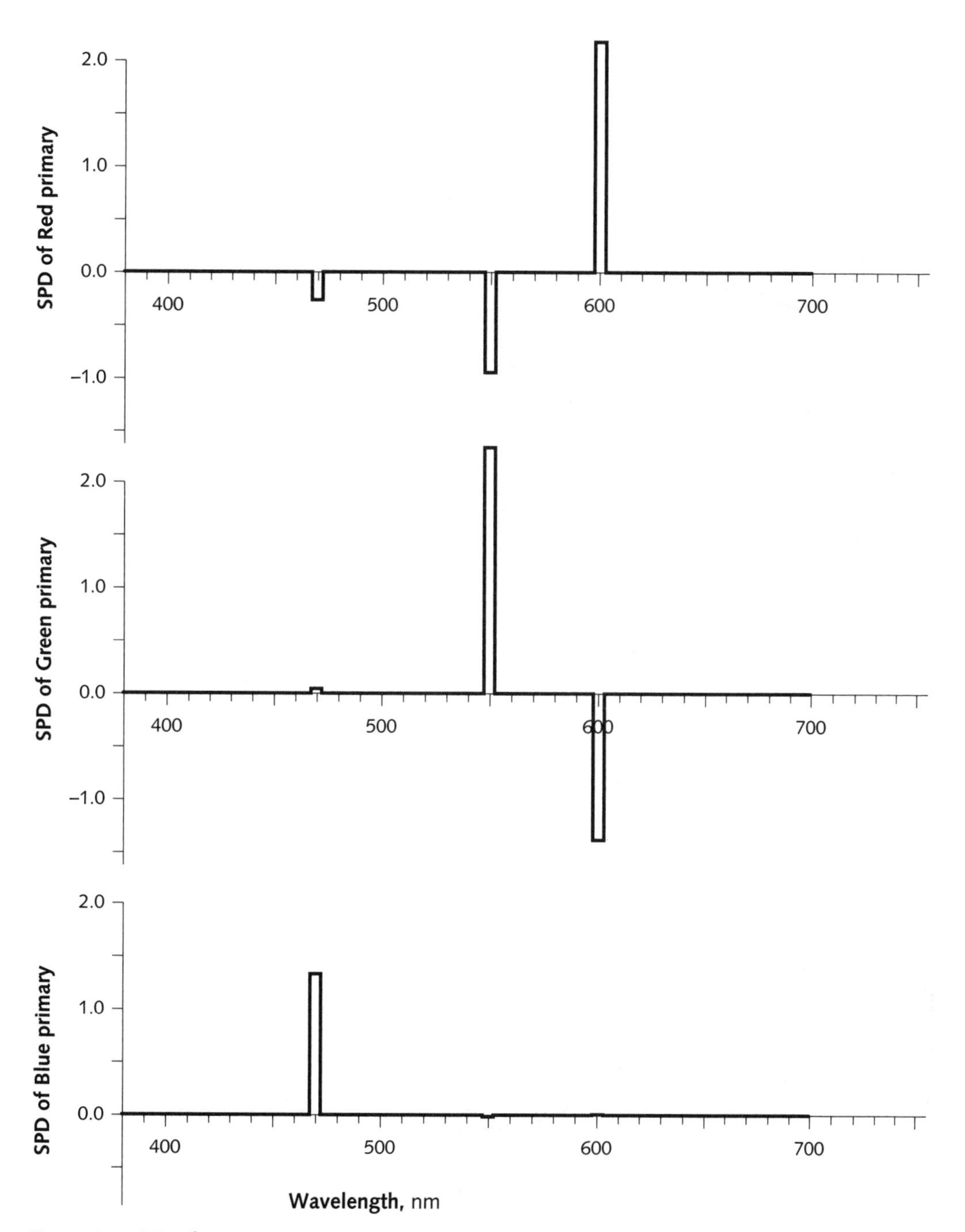

Figure 7.14 **SPDs for CIE XYZ primaries.** To directly reproduce a scene that has been analyzed using the CIE color matching functions requires *nonphysical* primaries having negative excursions, which cannot be realized in practice. Many different sets are possible. In this hypothetical example, the power in each primary is concentrated at the same three discrete wavelengths, 470, 550, and 600 nm.

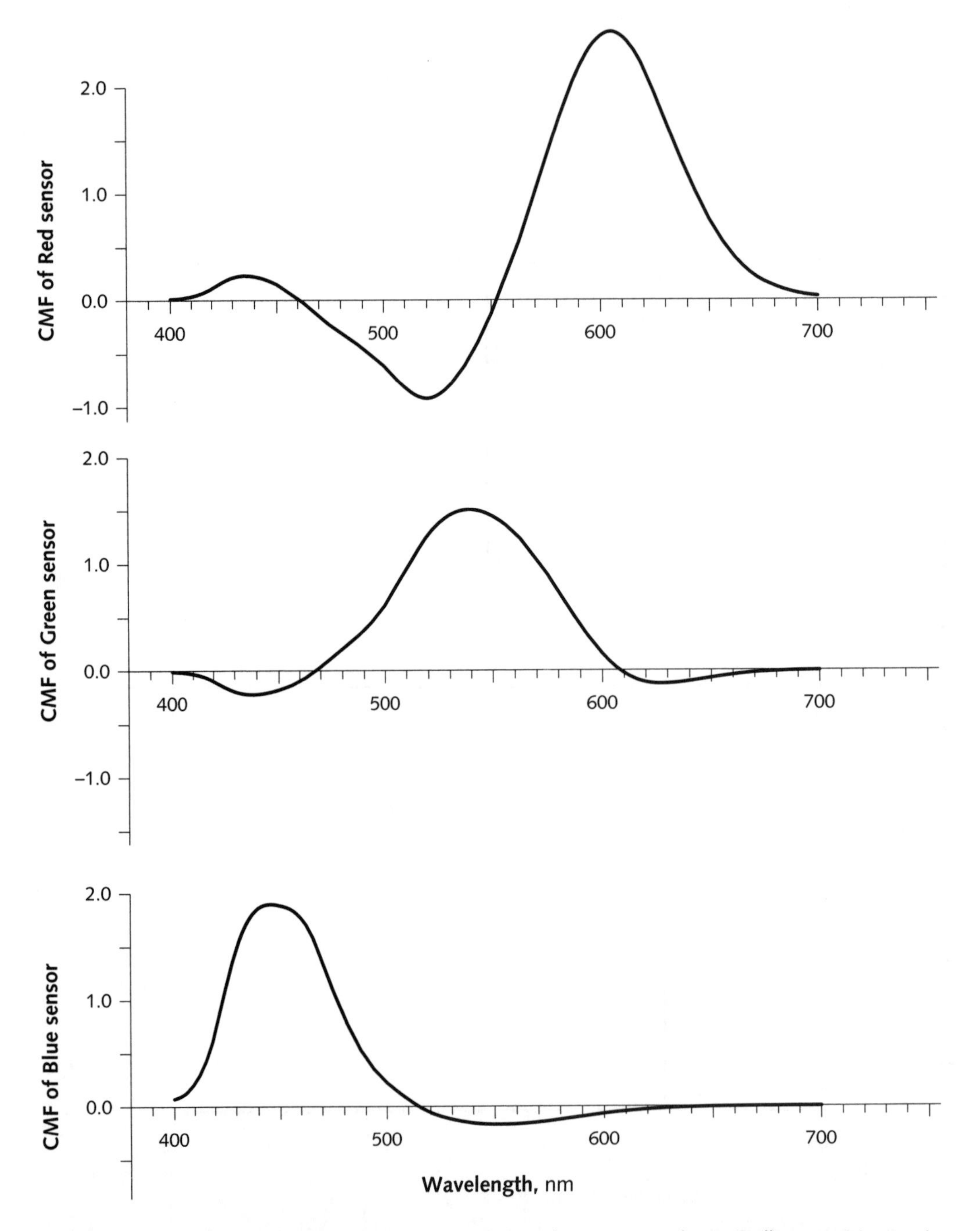

Figure 7.15 **CMFs for Rec. 709 primaries.** These analysis functions are theoretically correct to acquire *RGB* components for display using Rec. 709 primaries. The functions are not directly realizable in a camera or a scanner, due to their negative lobes. But they can be realized by a 3×3 matrix transformation of the CIE *XYZ* color matching functions of Figure 7.13.

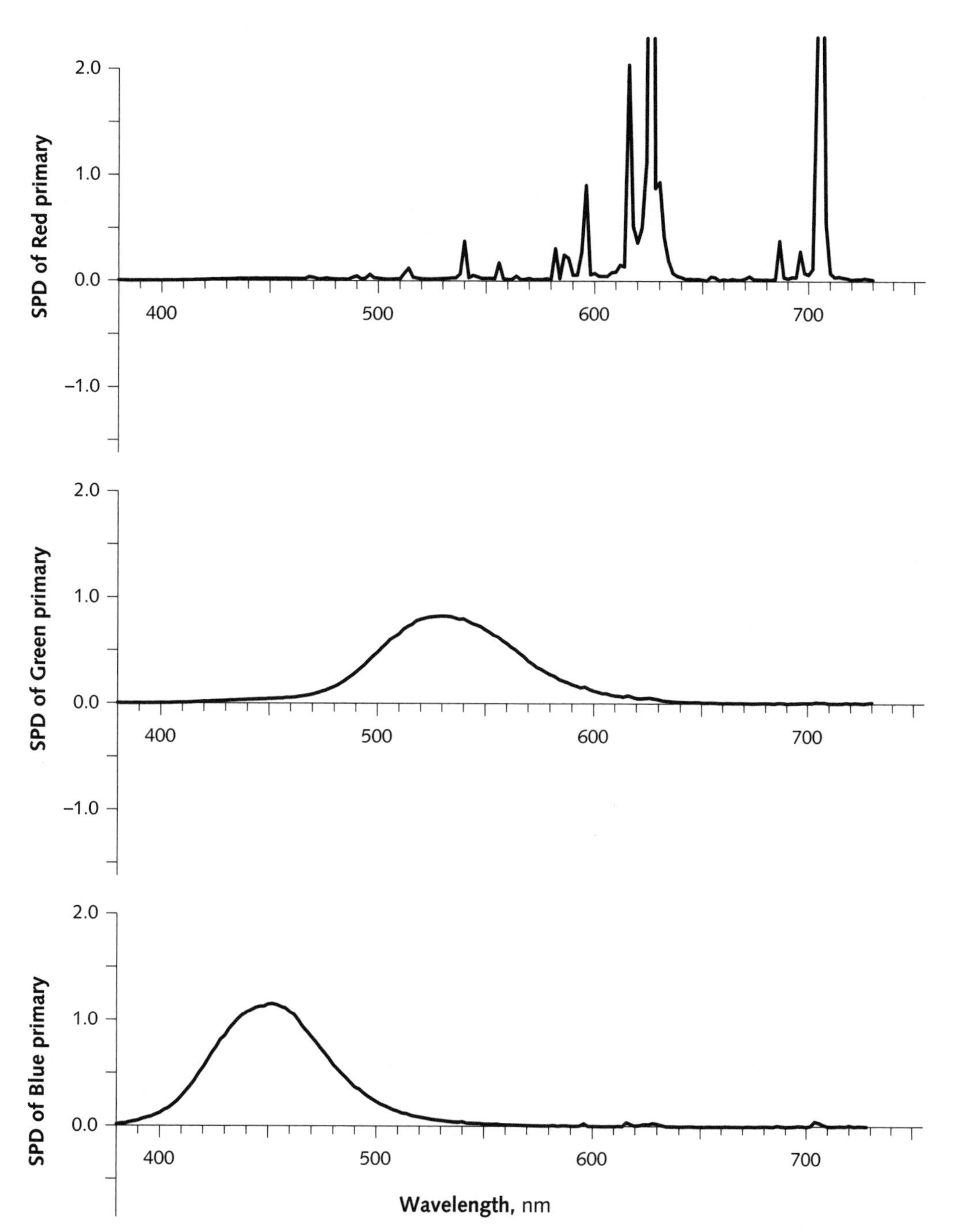

Figure 7.16 **SPDs for Rec. 709 primaries.** This set of SPDs has chromaticity coordinates that conform to SMPTE RP 145, similar to Rec. 709. Many SPDs could produce the same chromaticity coordinates; this particular set is produced by a Sony Trinitron monitor. The red primary uses *rare earth* phosphors that produce very narrow spectral distributions, different from the phosphors used for green or blue.

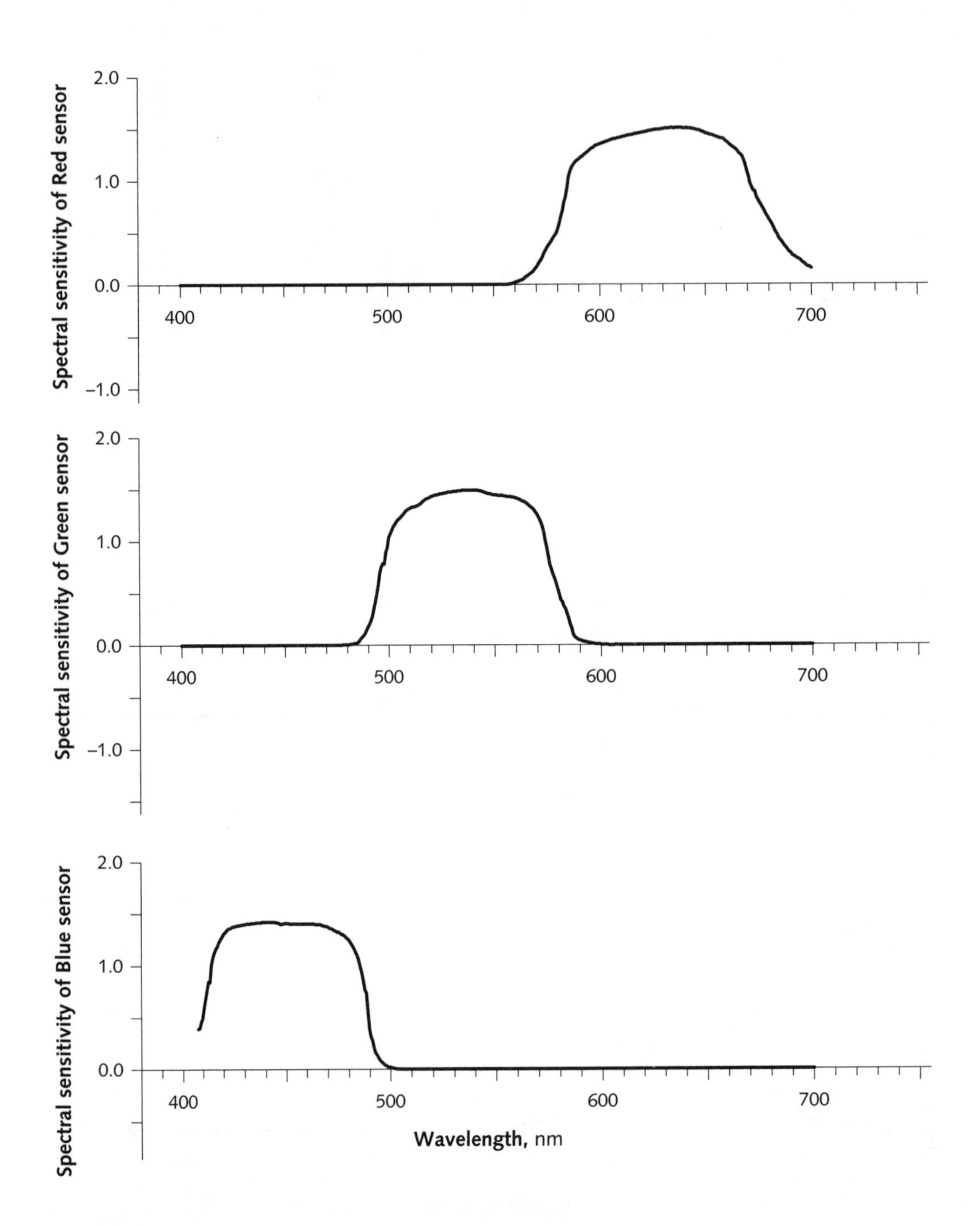

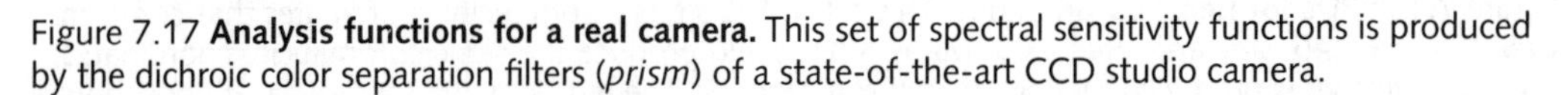
Figure 7.17 **Analysis functions for a real camera.** This set of spectral sensitivity functions is produced by the dichroic color separation filters (*prism*) of a state-of-the-art CCD studio camera.

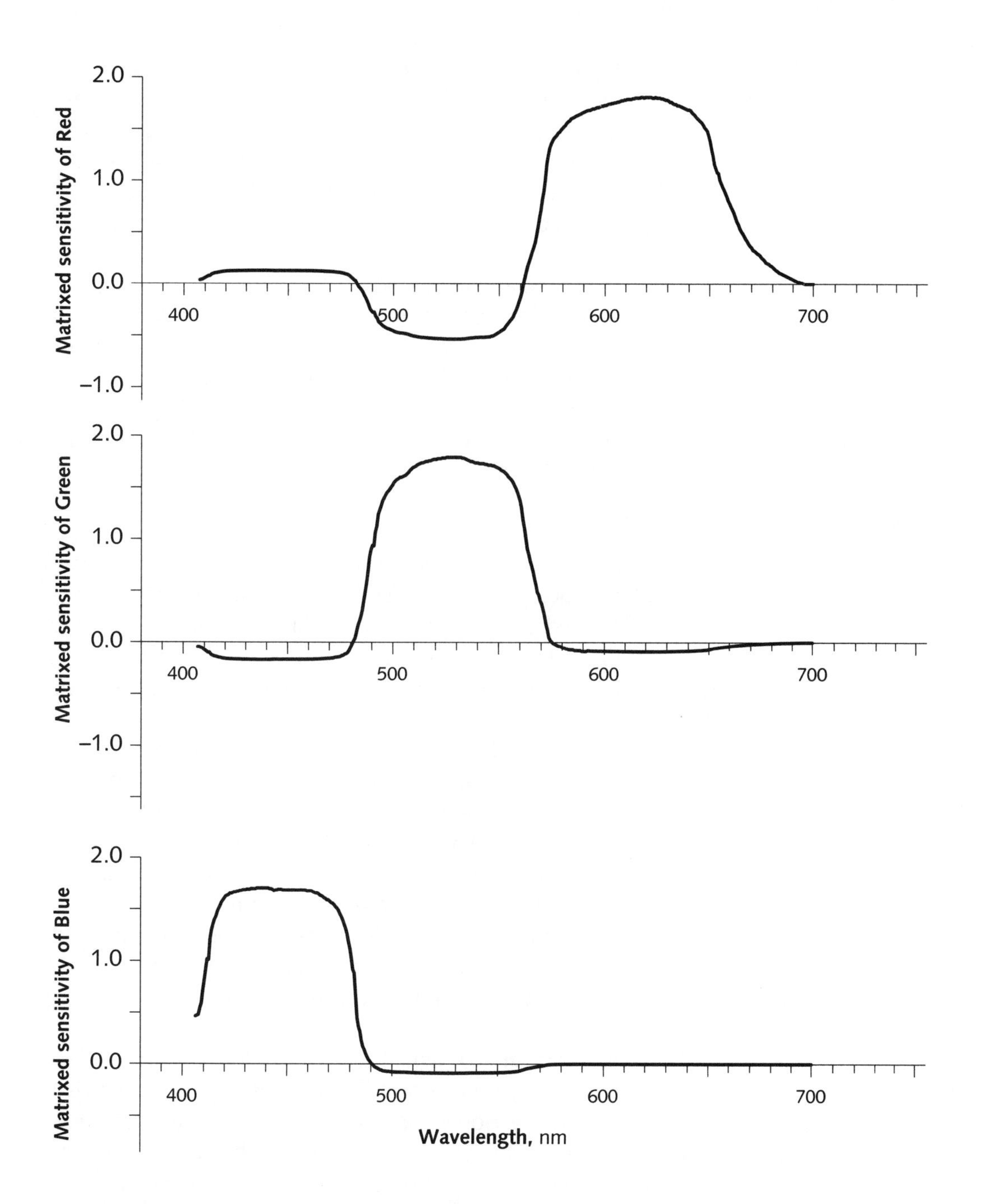

Figure 7.18 **CMFs of an actual camera after matrixing** for Rec. 709 primaries. These curves result from the analysis functions of Figure 7.17, opposite, being processed through a 3×3 matrix. Colors as "seen" by this camera will be accurate to the extent that these curves match the ideal CMFs for Rec. 709 primaries shown in Figure 7.15.

poor color reproduction will result if these signals are used directly to drive a display having Rec. 709 primaries. Figure 7.18 shows the same set of camera analysis functions processed through a 3×3 matrix transform. The transformed components will reproduce color more accurately.

In theory, and in practice, using a linear matrix to process the camera signals can reproduce color correctly. However, capturing all of the colors is seldom necessary in practice, as I will explain in the *Gamut* section below. Also, capturing the entire range of colors would incur a noise penalty, as I will describe in *Noise due to matrixing*, on page 147.

Gamut

Analyzing a scene with the CIE analysis functions produces distinct signals for all colors. But when transformed into components suitable for a set of physical display primaries, some of those colors – those whose chromaticity coordinates lie outside the triangle formed by the primaries – will have negative component values. In addition, colors outside the triangle of the primaries will have one or two primary components that exceed unity. These colors cannot be displayed correctly. Display devices usually clip signals that have negative values, and saturate signals whose values exceed unity. Visualized on the chromaticity diagram, a color outside the triangle of the primaries is reproduced at a point on the boundary of the triangle, on a line that joins the color with the white point.

If a scanner is designed to capture all colors, its complexity is necessarily higher and its performance is necessarily worse than a camera designed to capture a smaller range of colors. Thankfully, the range of colors encountered in the natural and man-made world is a small fraction of all of the colors. Although it is useful for an instrument such as a colorimeter to measure all colors, in an imaging system we are generally concerned with colors that occur frequently.

M. R. Pointer, "The Gamut of Real Surface Colours," *COLOR Research and Application*, vol. 5, no. 3 (Fall 1980).

M. R. Pointer, of Kodak Research Laboratories in England, characterized the distribution of frequently occurring *real surface colors*. The naturally occurring colors tend to lie in the central portion of the chromaticity diagram, where they can be encompassed by a well-chosen set of physical primaries. An imaging system performs well if it can display all or most of these colors.

Noise due to matrixing

Even if it were possible to display colors in the outer reaches of the chromaticity diagram, there would be a great practical disadvantage in doing so. Consider a camera that acquires *XYZ* tristimulus components, then transforms to Rec. 709 *RGB*:

Eq 7.9

$$\begin{bmatrix} R_{709} \\ G_{709} \\ B_{709} \end{bmatrix} = \begin{bmatrix} 3.240479 & -1.537150 & -0.498535 \\ -0.969256 & 1.875992 & 0.041556 \\ 0.055648 & -0.204043 & 1.057311 \end{bmatrix} \bullet \begin{bmatrix} X \\ Y \\ Z \end{bmatrix}$$

The coefficient 3.240479 in the upper left-hand corner determines the contribution from X at the camera into the red signal. An X component acquired with 10 mV of noise will inject 32.4 mV of noise into red: There is a noise penalty associated with the larger coefficients in the transform, and this penalty is quite significant in the design of a high quality camera.

Transformations between RGB and CIE XYZ

RGB values in a particular set of primaries can be transformed to and from CIE *XYZ* by a 3×3 matrix transform. These transforms involve *tristimulus values*, that is, sets of three linear-light components that conform to the CIE color matching functions. CIE *XYZ* is a special case of tristimulus values. In *XYZ*, any color is represented by an all-positive set of values. SMPTE has standardized a procedure for computing these transformations.

SMPTE RP 177-1993, *Derivation of Basic Television Color Equations.*

To transform from Rec. 709 *RGB* (with its D_{65} white point) into CIE *XYZ*, use the following transform. Because white is normalized to unity, the middle row sums to unity:

Eq 7.10

$$\begin{bmatrix} X \\ Y \\ Z \end{bmatrix} = \begin{bmatrix} 0.412453 & 0.357580 & 0.180423 \\ 0.212671 & 0.715160 & 0.072169 \\ 0.019334 & 0.119193 & 0.950227 \end{bmatrix} \bullet \begin{bmatrix} R_{709} \\ G_{709} \\ B_{709} \end{bmatrix}$$

The column vectors of this matrix are the *XYZ* tristimulus values of pure red, green, and blue. To recover primary chromaticities from such a matrix, compute little *x* and *y* for each *RGB* column vector. To recover the white point, transform RGB = [1, 1, 1] to *XYZ*, then compute *x* and *y*.

To transform from CIE *XYZ* into Rec. 709 *RGB*, use this transform:

Eq 7.11

$$\begin{bmatrix} R_{709} \\ G_{709} \\ B_{709} \end{bmatrix} = \begin{bmatrix} 3.240479 & -1.537150 & -0.498535 \\ -0.969256 & 1.875992 & 0.041556 \\ 0.055648 & -0.204043 & 1.057311 \end{bmatrix} \bullet \begin{bmatrix} X \\ Y \\ Z \end{bmatrix}$$

This matrix has some negative coefficients: *XYZ* colors that are *out of gamut* for Rec. 709 *RGB* transform to *RGB* components where one or more components are negative or greater than unity.

Transforms among RGB systems

K. Blair Benson, *Television Engineering Handbook, Featuring HDTV Systems, Revised Edition*, revised by Jerry C. Whitaker. New York: McGraw-Hill, 1992. This supersedes the Second Edition.

RGB values in a system employing one set of primaries can be transformed to another set by a 3×3 linear-light matrix transform. Generally these matrices are normalized for a white point luminance of unity. For details, see *Television Engineering Handbook*.

As an example, here is the transform from SMPTE RP 145 *RGB* to Rec. 709 *RGB*:

Eq 7.12

$$\begin{bmatrix} R_{709} \\ G_{709} \\ B_{709} \end{bmatrix} = \begin{bmatrix} 0.939555 & 0.050173 & 0.010272 \\ 0.017775 & 0.965795 & 0.016430 \\ -0.001622 & -0.004371 & 1.005993 \end{bmatrix} \bullet \begin{bmatrix} R_{145} \\ G_{145} \\ B_{145} \end{bmatrix}$$

To transform EBU 3213 RGB to Rec. 709:

Eq 7.13

$$\begin{bmatrix} R_{709} \\ G_{709} \\ B_{709} \end{bmatrix} = \begin{bmatrix} 1.044036 & -0.044036 & 0. \\ 0. & 1. & 0. \\ 0. & 0.011797 & 0.988203 \end{bmatrix} \bullet \begin{bmatrix} R_{EBU} \\ G_{EBU} \\ B_{EBU} \end{bmatrix}$$

Transforming among *RGB* systems may lead to an *out of gamut RGB* result, where one or more *RGB* components are negative or greater than unity.

These transformations produce accurate results only when applied to tristimulus (linear-light) components. To transform nonlinear *R'G'B'* requires an inverse transfer function to bring the components into the linear-light domain before the matrix multiplication, then reapplication of the transfer function. The transformation matrices of Equation 7.12 and Equation 7.13 are similar to the identity matrix: The terms on the diagonal are close to unity, and the terms off-diagonal are close to zero. In these cases, if the transform is computed in the nonlinear (gamma-corrected) *R'G'B'* domain, the resulting error will be small.

Camera white reference

There is an implicit assumption in television that the camera operates as if the scene were illuminated by a source having the chromaticity of CIE D_{65}. In practice, television studio lighting is often accomplished by tungsten lamps, and scene illumination is often deficient in the shortwave (blue) region of the spectrum. This situation is accommodated by adjusting the gain of the red, green, and blue components of the scene so that a white object reports the values that would be reported if the scene illumination had the same tristimulus values as CIE D_{65}.

Monitor white reference

In additive mixture the illumination of the reproduced image is generated entirely by the display device. In particular, reproduced white is determined by the characteristics of the display, and is not dependent on the environment in which the display is viewed. In a completely dark viewing environment such as a cinema theater, this is desirable. However, in an environment where the viewer's field of view encompasses objects other than the display, the viewer's notion of "white" is likely to be influenced or even dominated by what he or she perceives as "white" in the ambient. To avoid subjective mismatches, the chromaticity of white reproduced by the display and the chromaticity of white in the ambient should be reasonably close.

Modern blue CRT phosphors are more efficient with respect to human vision than red or green. In a quest for brightness at the expense of color accuracy, computer monitors commonly have a white point with a color temperature in the range of 9300 K, and produce a white having twice as much blue as the standard daylight reference CIE D_{65} used in television. This results in computer monitors and computer pictures that look excessively blue. This situation can be corrected by calibrating the monitor to a white reference with a lower color temperature.

Wide gamut reproduction

Charles A. Poynton, "Wide Gamut Device-Independent Colour Image Interchange," in *Proceedings of the International Broadcasting Convention*, 1994, IEE Conference Publication No. 397, 218–222.

For much of the history of color television, cameras were designed to incorporate assumptions about the color reproduction capabilities of color CRTs. But nowadays, video production equipment is being used to originate images for a wider range of applications than just television broadcast. The desire to make video cameras suitable for originating images for these wider applications has led to proposals for video coding to accommodate increased color range.

Scanning colored media

The *CIE XYZ tristimulus* section, on page 122, described the analysis filters necessary to obtain three tristimulus values from incident light from a colored object. But if the incident light has itself been formed by an imaging system based on three components (or *records*), the spectral composition of light from the original scene has already been analyzed. To perform a spectral analysis based on the CIE color matching functions is unnecessary at best, and is likely to introduce degraded color information. A much more direct, practical, and accurate approach is to perform an analysis using filters tuned to the spectrum of the colorants of the reproduced image, and thereby extract the three color records present in the reproduction being scanned.

In practical terms, if an image to be scanned is recorded on color film, or printed by color photography or offset printing, then better results will be obtained by using three narrowband filters designed to extract the color components that have already been recorded on the media. A *densitometer* is an instrument that performs this task. A densitometer produces not linear-light tristimulus values, but logarithmic *density* values. Several spectral sensitivity functions have been standardized for densitometers for use in different applications.

"Subtractive" mixture, CMY

I described how to produce a color by *Additive mixture, RGB* on page 132. Another way to produce a range of color mixtures is to selectively remove portions of the spectrum from a relatively broadband illuminant. This is illustrated in Figure 7.19 below.

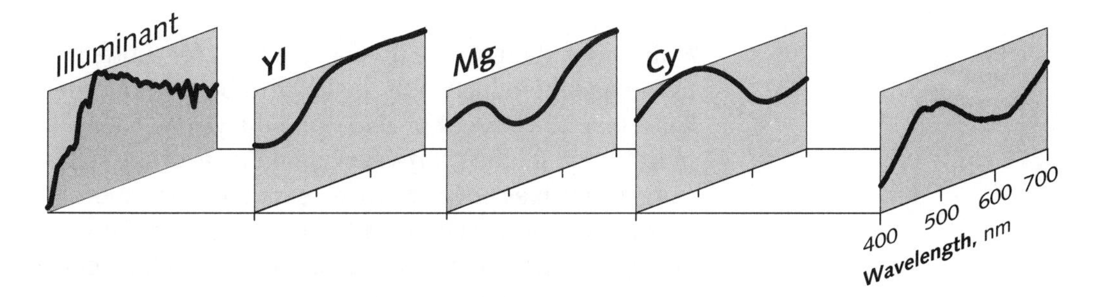

Figure 7.19 **Subtractive mixture,** as employed in color photography and color offset printing, is illustrated in this diagram. The colorants act in succession to remove spectral power from the illuminant. In physical terms, the spectral power of the mixture is, at each wavelength, the product of the spectrum of the illuminant and the transmission of the colorants: The mixture could be called *multiplicative*. If the amount of each colorant is represented in the form of *density* – the negative of the base 10 logarithm of transmission – then color mixtures can be determined by subtraction.

The illuminant at the left of the sketch produces light over most or all of the visible spectrum, and each successive filter transmits some portion of the band and attenuates other portions. The overall attenuation of each filter is controlled: In the case of color photography, this is accomplished by varying the thickness of each dye layer that functions as a filter, the shape of each curve remaining constant.

In physical terms, the spectrum of the mixture is the wavelength-by-wavelength product of the spectrum of the illuminant and the spectral transmission curves of each of the colorants. The spectral transmission curves of the colorants multiply, so this method of color reproduction should really be called *multiplicative*. Photographers and printers have for many decades measured transmission in base 10 logarithmic *density* units. Table 7.5, in the margin, shows some values. When a printer or photographer computes the effect of filters in tandem, she subtracts density values from unity, instead of multiplying transmission values together, so she calls the system *subtractive*.

Transmission	Density
1	0
0.1	1
0.01	2
0.001	3

Table 7.5 **Density examples.**

To achieve a wide range of colors in a subtractive system requires filters that appear colored cyan, yellow, and magenta (*CMY*). Figure 7.19 illustrates the situation of a transmissive display, such as in the projection of color photographic film.

RGB information can be used as the basis for subtractive image reproduction. If the color to be reproduced has a blue component of zero, then the yellow filter must attenuate the shortwave components of the spectrum as much as possible. To increase the amount of blue to be reproduced, the attenuation of the yellow filter should decrease. This reasoning leads to the "one minus *RGB*" relationships shown here:

$$
\begin{aligned}
Cy &= 1 - R \\
Mg &= 1 - G \\
Yl &= 1 - B
\end{aligned}
$$

Cyan in tandem with magenta produces blue, cyan with yellow produces green, and magenta with yellow produces red. Here is a memory aid:

R	*G*	*B*		*R*	*G*	*B*
Cy	*Mg*	*Yl*		*Cy*	*Mg*	*Yl*

Additive primaries are at the top, subtractive at the bottom. On the left, magenta and yellow filters combine subtractively to produce red. On the right, red and green sources combine additively to produce yellow. Here *RGB* and *CMY* are ordered as they are usually spoken and written; this is reversed from the order by increasing wavelength used in the graphs.

Unwanted absorptions

Although the "one-minus-*RGB*" relationships are a good starting point, they are not useful in practice. One problem is that the relationships assume both *RGB* and *CMY* to represent linear-light tristimulus values. But the quantities encountered in practice of are seldom linear-light. Another problem is that the relationships do not incorporate any information about tristimulus values of the assumed *RGB* primaries or the assumed *CMY* colorants: Obviously the spectral absorption of the colorants influences reproduced color, but the simplistic *one-minus-RGB* model embeds no information about the spectra!

The most serious practical problem with so-called subtractive color mixture is that any overlap among the absorption spectra of the colorants results in "unwanted absorption" in the mixture. For example, the magenta dye is intended to absorb only medium wavelengths, roughly between 500 nm and 600 nm. But for the actual dyes shown in Figure 7.19, magenta has considerable absorption at short wavelengths. Here, the amount of shortwave light that contributes to any mixture is not only a function of the amount of attenuation introduced by the yellow filter, but also the incidental attenuation introduced by the magenta filter.

There is no simple way to compensate these interactions, because the multiplication of spectra is not mathematically linear. In practice, the conversion from *RGB* to *CMY* is accomplished using either fairly complicated polynomial arithmetic, or by three-dimensional interpolation of lookup tables.

In a subtractive mixture, reproduced white is determined by characteristics of the colorants and by the spectrum of the illuminant at the display. In a reproduction such as a color photograph that is illuminated by the ambient light in the viewer's environment, mismatch between the white reference in the scene and the white reference in the viewing environment is eliminated. In subtractive reproduction, where the display generates its own illumination, the subjective effect of white mismatch may need to be considered.

Further reading

The bible of colorimetry is *Color Science*, by Wyszecki and Styles. But it's daunting. For a condensed version, read Judd and Wyszecki's *Color in Business, Science, and Industry*. It is directed to the color industry: ink, paint, and the like. For an approachable introduction to the same theory, accompanied by descriptions of image reproduction, consult Hunt's classic work.

For a discussion of nonlinear *RGB* in computer graphics, read Lindbloom's *SIGGRAPH* paper.

In a computer graphics system, once light is on its way to the eye, any tristimulus-based system can accurately represent color. But the interaction of light and objects involves spectra, not tristimulus values. In computer generated imagery (CGI), the calculations actually involve sampled SPDs, even if only three components are used. Roy Hall discusses these issues.

Günter Wyszecki and W. S. Styles, *Color Science: Concepts and Methods, Quantitative Data and Formulae, Second Edition.* New York: John Wiley & Sons, 1982.

Deane B. Judd and Günter Wyszecki, *Color in Business, Science, and Industry,* Third Edition. New York: John Wiley & Sons, 1975.

R. W. G. Hunt, *The Reproduction of Colour in Photography, Printing & Television*, Fifth Edition. Tolworth, England: Fountain Press, 1995.

Bruce Lindbloom, "Accurate Color Reproduction for Computer Graphics Applications," in *Computer Graphics*, vol. 23, no. 3 (July 1989), 117–126.

Roy Hall, *Illumination and Color in Computer Generated Imagery.* Berlin: Springer-Verlag, 1989.

Luma and color differences 8

This chapter describes color coding systems that are used to convey image data derived from additive (RGB) primaries. I outline nonlinear $R'G'B'$, explain the formation of *luma* Y' as a weighted sum of these nonlinear signals, and introduce the $B'-Y'$ and $R'-Y'$ *color difference* components.

The design of a video coding system is necessarily rooted in detailed knowledge of human color perception. However, once this knowledge is embodied in a coding system, what remains is physics, mathematics, and signal processing. This chapter concerns only the latter domains.

Color acuity

A monochrome video system should ideally sense luminance, described on page 82. Luminance should then be transformed by the gamma correction circuitry of the camera, as described in *Gamma in video*, on page 100, into a signal that takes into account the properties of lightness perception. At the receiver, the CRT itself has the required inverse transfer function.

To transmit a color image, three components – red, green, and blue, according to *Additive mixture, RGB*, on page 132 – could be acquired and transmitted. To minimize the visibility of noise or quantization, the *RGB* components should be coded nonlinearly.

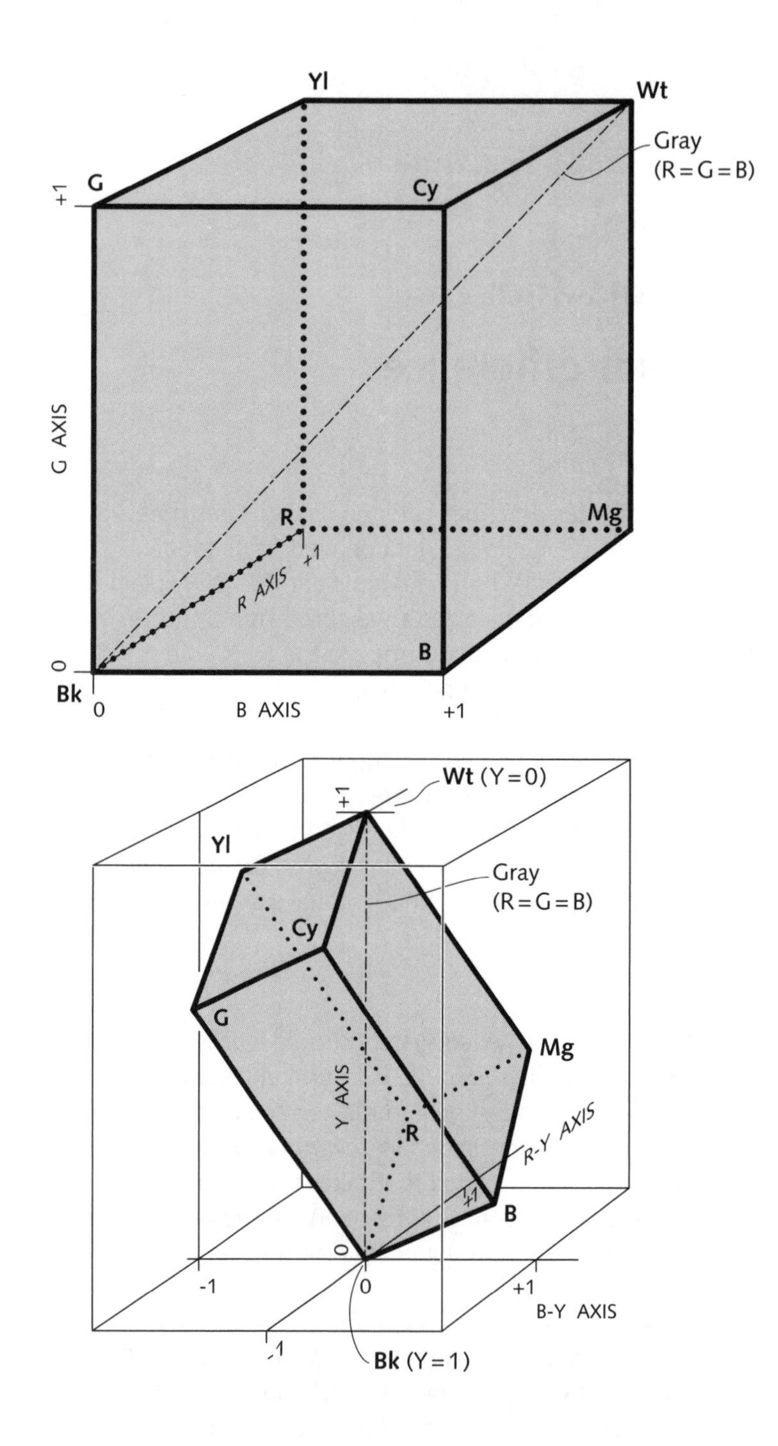

Figure 8.1 **RGB and Y, B–Y, R–Y cubes.**

RGB and Y, B–Y, R–Y color cubes

Red, green, and blue tristimulus primary components, as detailed in *Color science for video*, on page 115, can be considered as the coordinates of a three-dimensional color space. Coordinate values between zero and unity define the unit cube of this space, as sketched at the top of Figure 8.1 opposite.

The drawback of conveying *RGB* components of an image is that each component requires relatively high spatial resolution: Transmission or storage of a color image using *RGB* components requires a channel capacity three times that of a grayscale image. Human vision has considerably less spatial acuity for color information than for brightness. As a consequence of the poor color acuity of vision, a color image can be coded into a wideband component representative of brightness, and two narrowband color components where each color component has substantially less spatial resolution than brightness. In an analog system, each color channel can have less bandwidth, typically one-third that of brightness. In a digital system each color channel can have considerably less data rate (or data capacity) than brightness. There is evidence that the human visual system forms an achromatic channel and two chromatic color-difference channels at the retina.

Here the term *color difference* refers to a signal formed as the difference of two color components. In other contexts, the term can refer to a numerical measure of the perceptual distance between two colors.

Green dominates luminance: Between 60 and 70 percent of brightness luminance green information. Therefore it is sensible – and advantageous for signal-to-noise reasons – to base the color signals on the other two primaries. The simplest way to "remove" brightness from blue and red is to subtract it, to form a pair of *color difference* components.

At the bottom of Figure 8.1, the unit RGB cube is transformed into luminance *Y* and color difference components *B–Y* and *R–Y*. Once color difference signals have been formed, they can be subsampled to reduce bandwidth or data capacity, without the observer's noticing. Bandwidth reduction of analog color difference components is accomplished using analog lowpass filters that remove horizontal color detail. Digital video systems use digital filters that remove horizontal color detail, and possibly also remove vertical color detail. To get good results in digital video you should not simply drop and replicate pixels, but use proper decimation and interpolation filters designed according to the principles explained in *Filtering and sampling*, on page 43.

The unit *RGB* cube may be represented by three primary components of 8 bits each. When transformed to *Y, B−Y, R−Y* coordinates, the volume of the unit cube shrinks. If *Y, B−Y, R−Y* space is represented in three 8-bit coordinates, only one-quarter of those codes can be exercised by the codes within the unit RGB cube. *Y, B−Y, R−Y* is necessarily wasteful of codes, but this disadvantage is more than compensated by the opportunity to subsample.

The numerical values used in Figure 8.2, and subsequent figures in this chapter, reflect the Rec. 601 luma coefficients. For HDTV, the coefficients might be different. See *Rec. 601 luma*, on page 165.

Figure 8.1 is labeled in terms of linear-light (tristimulus) components. For reasons that will be described in the next several sections, it is standard in video to compute luma and color difference components from *nonlinear* primary *R′*, *G′*, and *B′* components, rather than tristimulus *R*, *G*, and *B* primary components. In this case the axes of Figure 8.1 are relabeled with primes: CIE luminance *Y* becomes video luma *Y′*. Figure 8.2 opposite shows three orthographic views of the nonlinear *R′G′B′* cube transformed into (*Y′*, *B′−Y′*, *R′−Y′*).

Constant luminance

Ideally, a color video system would operate similarly to a monochrome system. First, a true CIE luminance component *Y* would be formed as a weighted sum of *RGB* tristimulus components according to the principles of *Transformations between RGB and CIE XYZ* on page 147. Then, luminance would be subjected to a nonlinear transfer function, taking into account the lightness sensitivity of vision.

The nonlinear components representing color would ideally be formed by imposing a nonlinear transfer function on blue, and red, then subtracting nonlinearly

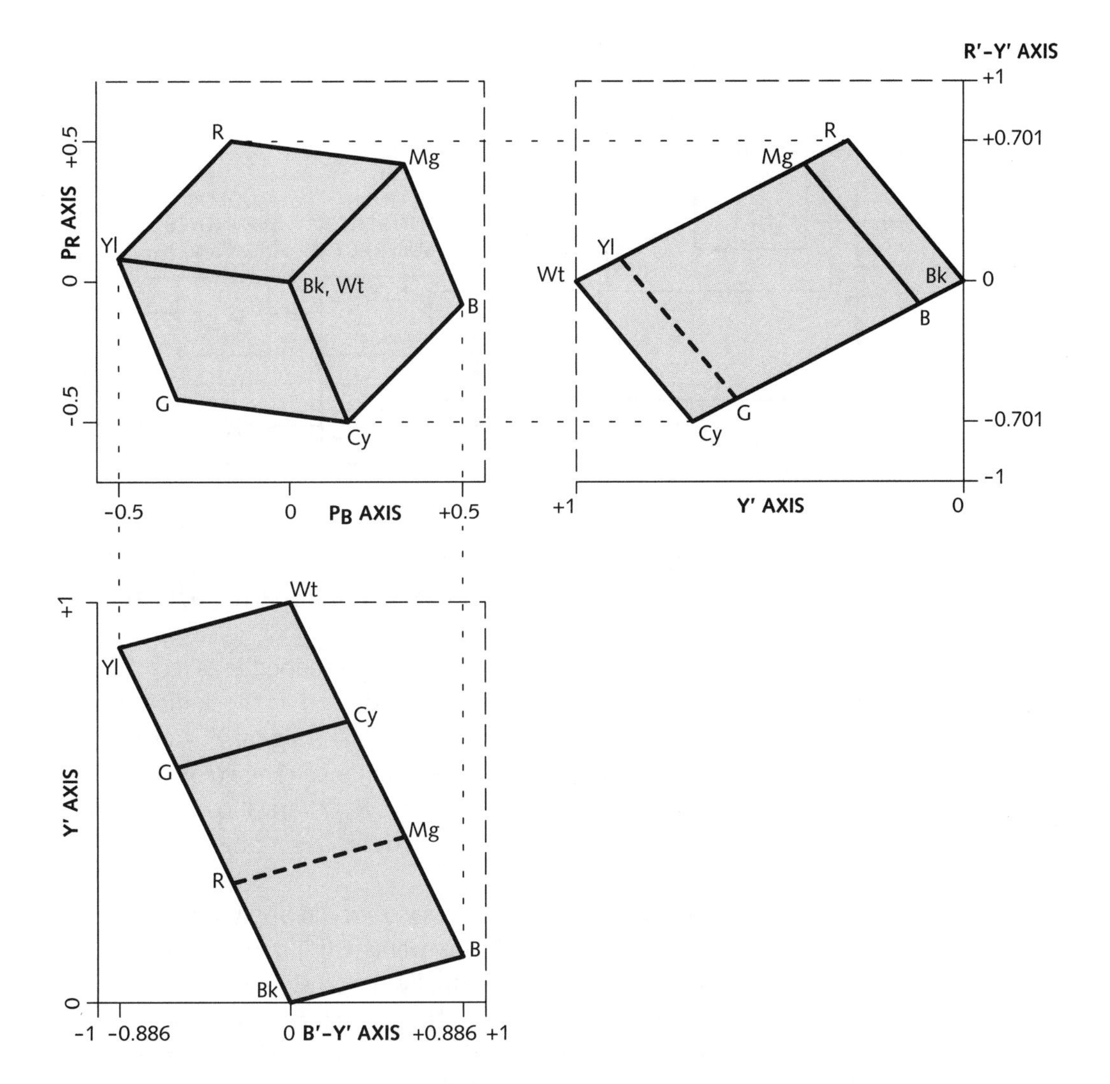

Figure 8.2 **Y', B'–Y', R'–Y' orthographic views.** The unit *R'G'B'* cube transformed into *Y', B'–Y', R'–Y'* coordinates reveals, at the upper left, the hexagonal boundary familiar to video engineers from vectorscope displays. The side view (*Y', R'–Y'*) to the right of the hexagon, and (*Y', B'–Y'*) below, are related to the *lightning* displays used in component video monitoring equipment. P_B and P_R components are also indicated; these components are scaled from *B'–Y'* and *R'–Y'* as will be detailed in *PBPR components*, on

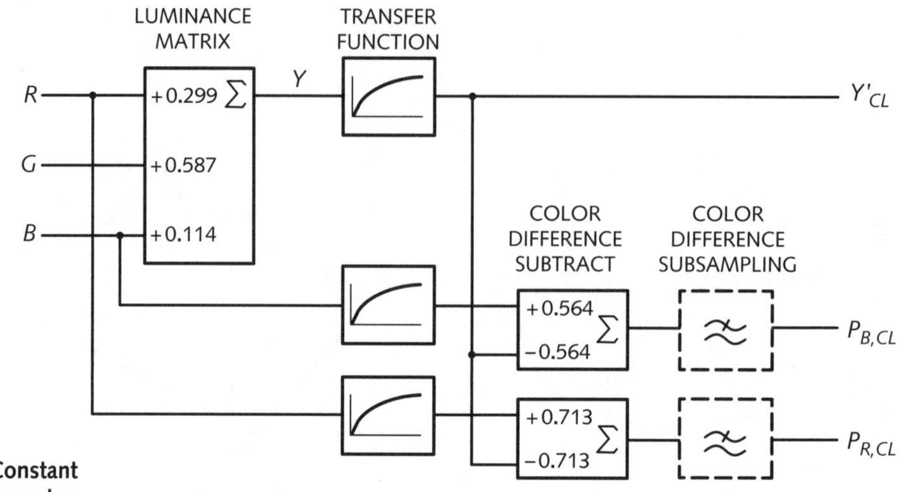

Figure 8.3 **Constant luminance encoder.**

coded luminance from each. The resulting color differences could be lowpass filtered – or, in a digital system, subsampled – to reduce the data capacity requirement.

The ideal scheme that I have described is called *constant luminance* encoding, and its block diagram is sketched in Figure 8.3 above. The signal Y'_{CL} represents nonlinearly coded luminance. The color difference components $B'-Y'$ and $R'-Y'$ are shown scaled to form P_B and P_R, as will be detailed on page 173.

$$\begin{bmatrix} 0 & 0 & 1 \\ 1.704 & -0.194 & -0.510 \\ 0 & 1 & 0 \end{bmatrix}$$

An ideal monitor or receiver would apply the inverse transfer function to the nonlinear luminance signal, to transform the perceptually coded signal back to luminance. Nonlinearly coded luminance would be added back to the color difference components to restore nonlinear blue and red primaries, and these would be subjected to the inverse transfer function to restore the linear blue and red components. Finally, the green tristimulus component is recovered as a weighted sum of luminance, blue, and red. The weighted sum is equivalent to the 3×3 matrix in the margin. A constant luminance decoder is sketched in Figure 8.4 opposite.

Unfortunately, the constant luminance scheme is impractical and uneconomical. A color CRT has three electron guns, each of which imposes a 2.5-power

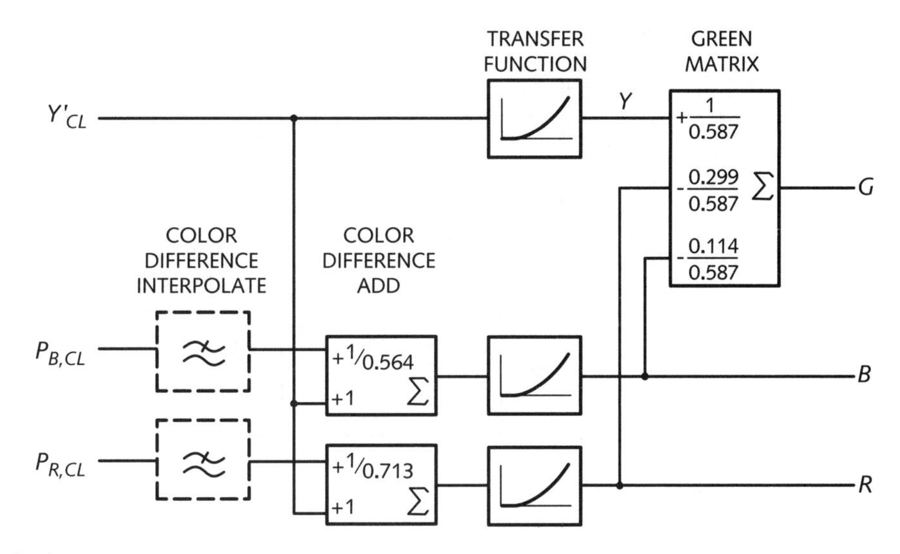

Figure 8.4 **Constant luminance decoder.**

function. The *RGB* tristimulus signals from this theoretical constant luminance decoder would have to be subjected to an inverse 0.4-power function prior to being presented to the CRT.

If the transfer functions in Figure 8.4 are interchanged with the green matrix – if green is recovered first, then the transfer function is applied – the transfer function can be accomplished within the electron guns of the CRT: No explicit components need to be provided. This reasoning led the NTSC in 1953 to adopt a *nonconstant luminance* design, which I will detail in a moment. The nonconstant luminance scheme has been adopted in all practical video systems, including NTSC, PAL, SECAM, component video, JPEG, MPEG, and HDTV. For a historical perspective, see Applebaum.

Sidney Applebaum, "Gamma Correction in Constant Luminance Color Television Systems," in *Proc. IRE*, October 1952, 1185–1195.

In discussions of new television standards, the suggestion is made periodically to include true constant luminance coding. However, the burden of reverse compatibility has so far proven insurmountable. Also, no conclusive scientific evidence proves the superiority of constant luminance coding, and the effect of subsampling of constant luminance color difference components has not been thoroughly studied.

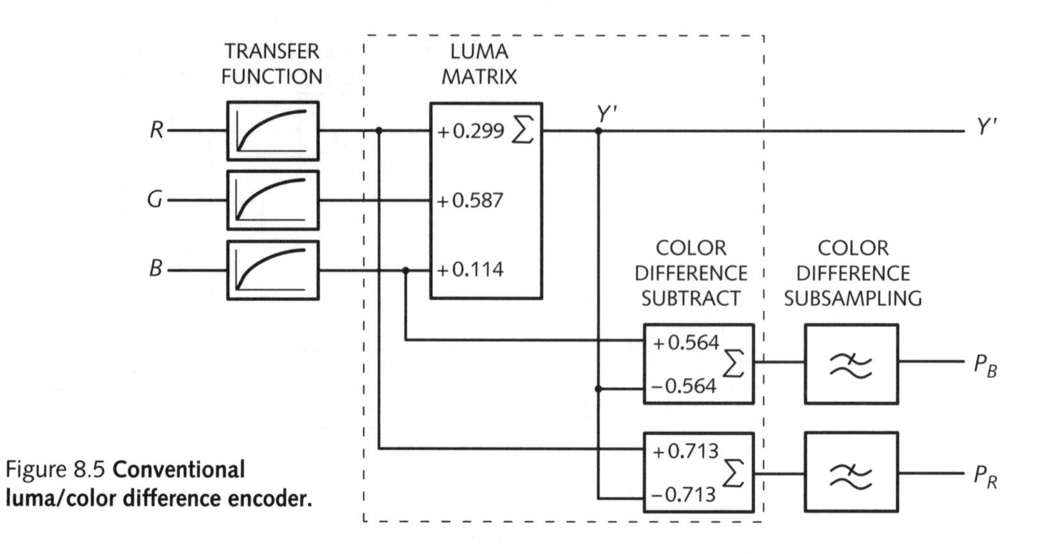

Figure 8.5 **Conventional luma/color difference encoder.**

Conventional luma/color difference coding

Constant luminance coding remains an intriguing possibility, but at present all video systems use nonconstant luminance coding, which I will now describe.

A conventional luma/color difference encoder is shown in Figure 8.5 above. First, a nonlinear transfer function is applied to the red, green, and blue linear (tristimulus) color components. Then video *luma* is formed as a weighted sum of gamma-corrected R', G', and B' primary components. B'-Y' and R'-Y' color difference components are formed by subtraction; in Figure 8.5, scaling to P_B and P_R components is indicated. Finally, the color difference components are lowpass filtered. In a digital system, the color difference components are subsampled, or *decimated* – this involves filtering, then discarding samples.

$$\begin{bmatrix} 0.299 & 0.587 & 0.114 \\ -0.169 & -0.331 & 0.5 \\ 0.5 & -0.419 & -0.081 \end{bmatrix}$$

In the rearranged block diagram of nonconstant luminance encoding, the weighted sum that forms luma and the pair of color difference subtractors – grouped together by dashed lines in the block diagram – are equivalent to a 3×3 matrix transform. The coefficients of the matrix are shown in the margin. Color coding equations and matrices will be detailed in *Component video color coding*, on page 171.

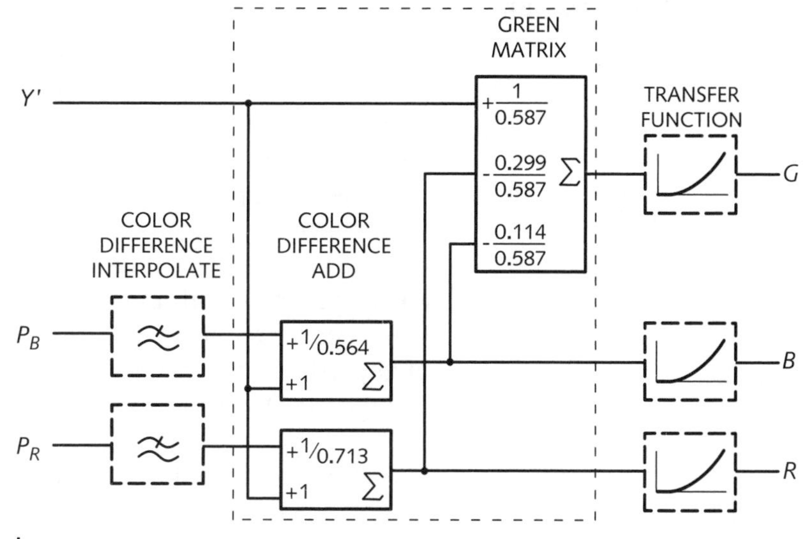

Figure 8.6 **Conventional luma/color difference decoder.**

Figure 8.6 above illustrates a conventional, *non-constant luminance* luma/color difference decoder. In a digital decoder, the omitted color difference components are interpolated. Luma is added to the color difference components to recover nonlinear blue and red components. A weighted sum of luma, blue, and red is formed to recover the nonlinear green component. Finally, all three components are subject to the inverse transfer function.

If the decoder is associated with a CRT display device, the CRT will impose the required transfer function, so it need not be explicitly computed.

$$\begin{bmatrix} 1 & 0 & 1.402 \\ 1 & -0.344 & -0.714 \\ 1 & 1.772 & 0 \end{bmatrix}$$

The blue and red color difference adders, and the weighted adder that recovers green, can all be considered together as a 3×3 matrix whose coefficients are indicated in the margin.

Now that I have explained the reasons for the adoption of this block diagram, I will detail the steps involved in the encoding of luma and color difference signals, starting with the formation of nonlinear R', G', and B' primary components.

Nonlinear Red, Green, Blue (R'G'B')

Video originates with linear-light (*tristimulus*) *RGB* primary components, usually represented in abstract terms in the range 0 (black) to +1 (white). In order to meaningfully determine a color from an *RGB* triple, the colorimetric properties of the primaries and the assumed white point – such as their CIE *xy* chromaticity coordinates – must be known. Colorimetric properties of a set of *RGB* components are discussed in *Color science for video*, on page 115. In the absence of any specific information, use the Rec. 709 primaries and the CIE D_{65} white point.

In *Gamma*, on page 91, I described how brightness information is coded nonlinearly, in order to achieve good perceptual performance from a limited number of bits. In a color system, the nonlinear transfer function described in that chapter is applied individually to each of the three *RGB* tristimulus components: From the set of *RGB* tristimulus (linear-light) values, three gamma-corrected primary signals are computed. Each is approximately the 0.45-power of the corresponding tristimulus value, similar to a square-root function.

Television standards documents have historically used the prime symbol (') – often combined with the letter *E* for voltage – to denote a component that incorporates gamma correction. For example, E'_R historically denoted the gamma-corrected red channel. Gamma correction is nowadays so taken for granted in video that the *E* and the prime symbol are usually dropped. This has led to considerable confusion among people attempting to utilize video technology in other domains.

I detailed the *Rec. 709 transfer function* on page 102. Although standardized for HDTV, it is now applied to conventional video as well. Expressed in three components, for tristimulus values greater than a few percent:

Eq 8.1

$$\begin{aligned} R'_{709} &= 1.099 R^{0.45} - 0.099 \\ G'_{709} &= 1.099 G^{0.45} - 0.099 \\ B'_{709} &= 1.099 B^{0.45} - 0.099 \end{aligned}$$

SMPTE 240M for 1125/60 HDTV uses numerical values slightly different from those of Rec. 709. For tristimulus values greater than a few percent:

Eq 8.2

$$\begin{aligned} R'_{240M} &= 1.1115 R^{0.45} - 0.0228 \\ G'_{240M} &= 1.1115 G^{0.45} - 0.0228 \\ B'_{240M} &= 1.1115 B^{0.45} - 0.0228 \end{aligned}$$

Rec. 601 luma

The following luma equation is standardized for conventional television in Rec. 601, and applies to 525/59.94 video, 625/50 video, and PhotoYCC:

Eq 8.3

$$Y'_{601} = 0.299\, R' + 0.587\, G' + 0.114\, B'$$

As mentioned a moment ago, the *E* and prime symbols originally used for video signals have been elided over the course of time, and this has led to ambiguity of the *Y* symbol between color science and television.

The coefficients in the luma equation correspond to the sensitivity of human vision to each of the *RGB* primaries standardized for the coding. The low value of the blue coefficient is a consequence of saturated blue colors being perceived as having low brightness. The luma coefficients are also a function of the white point, or more properly, the *chromaticity of reference white*.

In theory, the coefficients in the luma equation should be derived from the primary and white chromaticities. The Rec. 601 luma coefficients stem from the original NTSC primaries standardized by the FCC in 1953. Although television primaries have changed over the years since the adoption of NTSC, and the primaries in use for 525/59.94 are the SMPTE RP 145 primaries, the coefficients of the luma equation for 525/59.94 and 625/50 video have remained unchanged. Consequently, the luma coefficients are no longer theoretically correct for the primaries now in use. However, this is of no practical significance except in very high quality conversion of signals among systems.

SMPTE 240M luma

SMPTE 240M standardizes luma coefficients for 1125/60, 1920×1035 HDTV that are "theoretically correct" for the SMPTE RP 145 primaries:

Eq 8.4

$$Y'_{240M} = 0.212\,R' + 0.701\,G' + 0.087\,B'$$

The SMPTE 240M coefficients are used in virtually all of the HDTV equipment that is deployed as I write this.

Rec. 709 luma

International agreement on Rec. 709 was achieved in 1990 on the basis of "theoretically correct" luma coefficients derived from the Rec. 709 primaries:

Eq 8.5

$$Y'_{709} = 0.2125\,R' + 0.7154\,G' + 0.0721\,B'$$

The difference between the 240M coefficients and the Rec. 709 coefficients is negligible for real images. However, test signals produce different numerical results for the two sets. Manufacturers and users have adopted the long-term goal of a transition to the Rec. 709 *primaries*. Whether there will be a transition to Rec. 709 *luma coefficients* is uncertain.

Errors due to nonconstant luminance

In an ideal constant luminance system, all true CIE luminance information would be conveyed in the Y' component. In a nonconstant luminance system, some brightness information necessarily "leaks" into the color difference components, the leakage being larger the further a color lies from the gray axis of the color cube. For colors near gray, leakage is minimized by using luma coefficients that are numerically identical to the coefficients that would be used to compute true CIE luminance. In other words, the correct luminance coefficients are used, but in the wrong equation!

If the color difference components are conveyed with the same bandwidth as luma, then this issue of "leakage" is of little concern. But the reason to code into color difference components is to reduce the bandwidth of the color information. When this is done, the discarding of high spatial frequencies also discards high-frequency luminance "leakage" components. When the image is reconstructed as RGB, the luminance that is reconstructed will be incorrect. The most obvious manifestation of this problem is at the green-to-magenta transition of the color bar test pattern – the dark band that appears at the center of the transition is a result of luminance being reproduced too low.

It has not been shown that using the "theoretically correct" coefficients delivers any practical advantage, but there is a huge advantage in maintaining a single standard set of luma coefficients. Until about 1985, television equipment was specialized for particular scanning standards, so differences in parameters between systems have not been important. But it is no longer economically feasible to design television equipment for just a continental market. If different markets standardize different parameter values that have no functional advantage, the complexity and cost of equipment rises for no good reason. Furthermore, the computing and communications industries find it difficult to imagine why a 640×480 image should be coded with one set of luma coefficients and a 1920×1080 image should be coded with a different set. The PhotoYCC color coding system of Kodak's PhotoCD system adopts the Rec. 601 luma coefficients, and Rec. 601 coefficients have been adopted for 1250/50 HDTV according to the latest revision of Rec. 709.

As I write this, there is a proposal to retain the Rec. 601 coefficients for all variants of HDTV.

Subsampling

The purpose of color difference coding is to reduce the information capacity for color information. The bandwidth of analog color difference components can be reduced by analog filtering. The data capacity of digital color difference components can be reduced by subsampling. Figure 1.14, *Chroma subsampling*, on page 25, sketches several digital subsampling schemes.

In 4:2:2 subsampling, alternate color difference samples are discarded at the encoder. At the decoder, the missing samples are approximated by interpolation. Obviously some color detail is lost, but this loss cannot be detected by a viewer at normal viewing distance.

Casual discarding of alternate samples can be viewed as point sampling; the operation runs the risk of introducing aliases. Aliasing can be minimized by filtering prior to discarding samples, according to the principles discussed in *Filtering and sampling*, on page 43.

Luma/color difference summary

If luma and color difference coding to be used for image interchange, it is important for the characteristics of red, green, and blue to be maintained from the input of the encoder to the output of the decoder. The chromaticities of the red, green, and blue primaries have not been mentioned in this chapter, but an implicit assumption has been made that the characteristics of the primaries match from across the whole system. The primaries upon which luma and color difference coding are based are known as the *interchange* (or *transmission*) *primaries*.

In practice, an encoder may be presented with *RGB* components whose chromaticities do not match the interchange primaries. In this case it may be necessary to insert, in front of the encoder, a 3×3 matrix to transform from the image capture primaries into the interchange primaries. Similarly, a decoder may be required to drive a display whose primaries are different from the interchange primaries: at the output of the

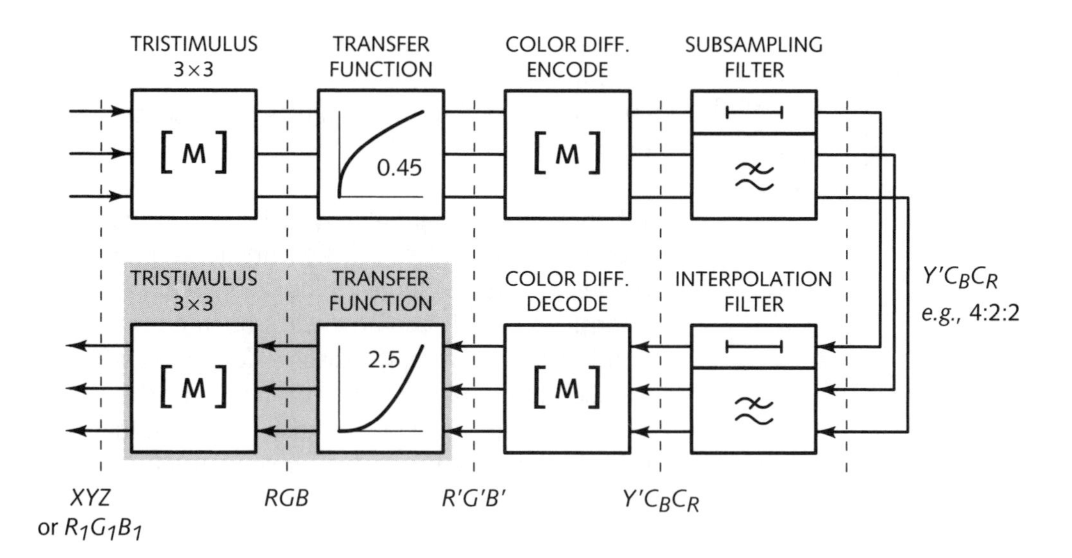

Figure 8.7 **Color difference encoder/decoder.**

decoder, in may be necessary to insert a 3×3 matrix to transform from the interchange primaries into the image display primaries.

Figure 8.7 above summarizes luma/color difference encoding and decoding. From linear *XYZ* – or linear $R_1G_1B_1$ whose chromaticities differ from the interchange standard – apply a 3×3 matrix transform to obtain linear *RGB* according to the interchange primaries. Apply a nonlinear transfer function (gamma correction) to each of the components to obtain nonlinear *R'G'B'*. Apply a 3×3 matrix to obtain luma and color difference components, such as $Y'P_BP_R$, $Y'C_BC_R$, or PhotoYCC. If necessary, apply a subsampling filter to obtain subsampled color difference components.

A decoder uses the inverse operations of the encoder, in the opposite order. The interpolation filter reconstructs wideband color difference components; in an analog decoder no explicit operation is needed. The 3×3 color difference matrix reconstructs nonlinear red, green, and blue primary components. The transfer functions restore the primary components to their

Figure 8.7 shows 3×3 matrix transforms being used for two distinctly different tasks. When someone hands you a 3×3, you have to ask whether it is intended for a linear or nonlinear task.

linear-light tristimulus values. Finally, the tristimulus 3×3 matrix transforms from the primaries of the interchange standard to the primaries implemented in the display device.

When a decoder is intimately associated with a CRT monitor, the decoder's transfer function is performed by the nonlinear voltage-to-intensity relationship intrinsic to the CRT, so no explicit operations are required for this step. If a display device has a transfer function that is different from a CRT, then this decoding step can apply a transfer function that is the composition of a 2.5-power function and the inverse transfer function of the display device.

If the display primaries match the interchange primaries, the decoder's 3×3 tristimulus matrix is not needed. If a CRT display has primaries not too different from the interchange primaries, then it may be possible to compensate the primaries by applying a 3×3 matrix in the nonlinear signal domain. But if the primaries are quite different, it will be necessary to make apply the transform between primaries in the tristimulus domain; see *Transforms among RGB systems*, on page 148.

This chapter outlined the principle of luma/color difference coding, and detailed the formation of luma according to several standards. Several different color difference component sets, differing mainly in scaling, are in use; the following chapter provides details.

Component video color coding 9

Chapter 8 explained color difference coding. Various scale factors are applied to the basic color difference components $B'-Y'$ and $R'-Y'$ for different applications. $Y'P_BP_R$ scaling is appropriate for component analog video such as BetaCam and M-II. $Y'C_BC_R$ scaling is appropriate for component digital video such as studio video, JPEG, and MPEG. Kodak's PhotoYCC uses scale factors optimized for the gamut of film colors.

Video uses the symbols U and V to represent certain color difference components. The CIE defines the pairs (u, v), (u', v'), and (u^*, v^*). All of these pairs represent *chromatic* or chroma information, but they are all numerically and functionally different. Video UV components are neither directly based on, nor superseded by, any of the CIE color spaces.

Technically, $Y'UV$ and $Y'IQ$ are component codes, so I describe them in this chapter; but both are intermediate quantities toward the formation of composite NTSC, PAL, and S-video. Neither is appropriate when the components are kept separate. Unfortunately the $Y'UV$ nomenclature is now used rather loosely, and it sometimes denotes any scaling of $B'-Y'$ and $R'-Y'$. $Y'IQ$ coding is obsolete. Neither $Y'UV$ nor $Y'IQ$ has a standard component interface.

The subscripts in C_BC_R and P_BP_R are often written in lowercase. I find this to compromise readability, so I write them in uppercase. Authors with great attention to detail sometimes "prime" these quantities to indicate their nonlinear nature, but because no practical image coding system employs linear color differences, I consider it safe to omit the primes.

The coding systems described in this chapter can be applied to various RGB primary sets – EBU 3213, SMPTE RP 145, or Rec. 709. See page 133.

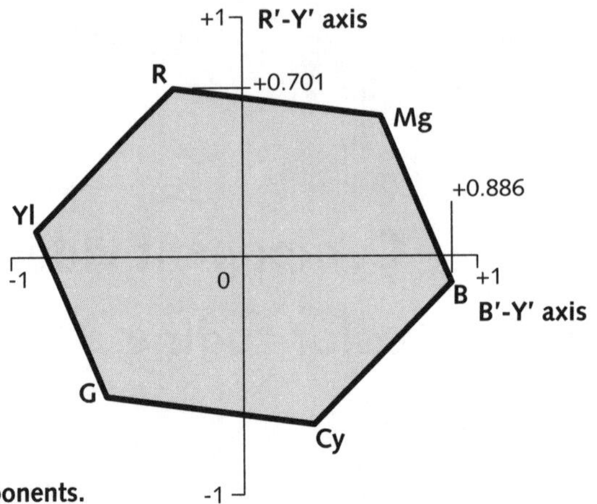

Figure 9.1 **B'–Y', R'–Y' components.**

B'–Y', R'–Y' components

As described on page 165, the Rec. 601 luma coefficients are used for conventional 525/59.94 and 625/50 video. With these coefficients, the $B'-Y'$ signal reaches its positive maximum at blue ($R=0$, $G=0$, $B=1$; $Y=0.114$; $B'-Y'=+0.886$) and its negative maximum at yellow ($B'-Y'=-0.886$). Analogously, the extrema of $R'-Y'$ occur at red and cyan at values ±0.701. These are inconvenient values for both digital and analog systems. The systems $Y'P_BP_R$, $Y'C_BC_R$, PhotoYCC, and $Y'UV$, to be described, all employ versions of (Y', $B'-Y'$, $R'-Y'$) that are scaled to place the extrema of the component signals at more convenient values.

To obtain (Y', $B'-Y'$, $R'-Y'$), from $R'G'B'$, for Rec. 601 luma, use this matrix transform:

Eq 9.1

$$\begin{bmatrix} Y'_{601} \\ B'-Y'_{601} \\ R'-Y'_{601} \end{bmatrix} = \begin{bmatrix} 0.299 & 0.587 & 0.114 \\ -0.299 & -0.587 & 0.886 \\ 0.701 & -0.587 & -0.114 \end{bmatrix} \bullet \begin{bmatrix} R' \\ G' \\ B' \end{bmatrix}$$

The numerical values used here, and in subsequent sections of this chapter, are based on the Rec. 601 luma coefficients. For HDTV, the coefficients might be different. See *Rec. 601 luma*, on page 165.

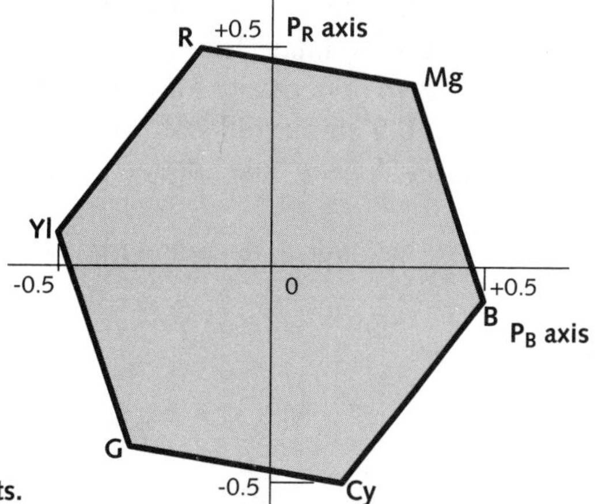

Figure 9.2 **$P_B P_R$ components.**

$P_B P_R$ components

If two color difference components are to be formed having identical unity excursions, then P_B and P_R color difference components are used. For Rec. 601 luma, the equations are these:

Eq 9.2

$$P_B = \frac{0.5}{1-0.114}\left(B' - Y'_{601}\right)$$
$$P_R = \frac{0.5}{1-0.299}\left(R' - Y'_{601}\right)$$

These scale factors, sometimes written 0.564 and 0.713, are chosen to limit the excursion of *each* color difference component to the range -0.5 to +0.5 with respect to unity luma excursion: 0.114 in the first expression above is the luma coefficient of blue, and 0.299 in the second is for red. In the analog domain, luma often ranges from 0 mV (black) to 700 mV (white), and P_B and P_R analog signals range ±350 mV.

Expressed in matrix form, the B'-Y' and R'-Y' rows of Equation 9.1 are scaled by $^{0.5}/_{0.886}$ and $^{0.5}/_{0.701}$. To encode from $R'G'B'$ where reference black is zero and reference white is unity:

Eq 9.3

$$\begin{bmatrix} Y'_{601} \\ P_B \\ P_R \end{bmatrix} = \begin{bmatrix} 0.299 & 0.587 & 0.114 \\ -0.168736 & -0.331264 & 0.5 \\ 0.5 & -0.418688 & -0.081312 \end{bmatrix} \bullet \begin{bmatrix} R' \\ G' \\ B' \end{bmatrix}$$

The first row comprises the luma coefficients; these sum to unity. The second and third rows each sum to zero, a necessity for color difference components. The +0.5 entries reflect the maximum excursion of P_B and P_R, for the blue and red primaries [0, 0, 1] and [1, 0, 0].

The inverse, decoding matrix is this:

Eq 9.4

$$\begin{bmatrix} R' \\ G' \\ B' \end{bmatrix} = \begin{bmatrix} 1 & 0 & 1.402 \\ 1 & -0.344136 & -0.714136 \\ 1 & 1.772 & 0 \end{bmatrix} \bullet \begin{bmatrix} Y'_{601} \\ P_B \\ P_R \end{bmatrix}$$

See *Component analog $Y'P_BP_R$ interface, SMPTE*, on page 219. Betacam and M-II equipment use nonstandard levels; see *Component analog $Y'P_BP_R$ interface, industry standard*, on page 220.

$Y'P_BP_R$ is employed by 525/59.94 and 625/50 component analog video equipment such as M-II and BetaCam, where P_B and P_R are conveyed with half the bandwidth of luma. For analog HDTV, the P_BP_R scaling is based on either SMPTE 240M or Rec. 709 luma coefficients, as described on page 166.

C_BC_R components

Rec. ITU-R BT.601-4 is the international standard for studio-quality component digital video. Luma is coded in 8 bits. Y' has an excursion of 219 and an offset of +16: Black is at code 16 and white is at code 235. Color differences C_B and C_R are coded in 8-bit two's complement form, with excursions of ±112 and offset of +128, for a range of 16 through 240 inclusive. The extremes of the coding range provide signal processing headroom and footroom, including the accommodation of ringing from analog and digital filters.

$Y'C_BC_R$ coding has a slightly smaller excursion for luma than for chroma: Luma has 219 "risers" compared to 224 for C_B and C_R. The notation C_BC_R distinguishes this set from P_BP_R, where the luma and chroma excursions are identical.

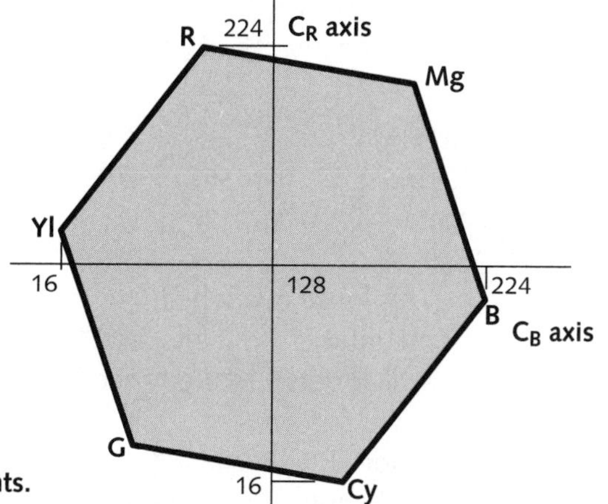

Figure 9.3 **$C_B C_R$ components.**

Expressed a set of equations in terms of $B'-Y'$ and $R'-Y'$, for Rec. 601 coding in 8 bits per component:

Eq 9.5

$$
\begin{aligned}
Y'_{219} &= 16 + 219Y' \\
C_B &= 128 + 112\left(\frac{0.5}{1-0.114}\left(B' - Y'_{601}\right)\right) \\
C_R &= 128 + 112\left(\frac{0.5}{1-0.299}\left(R' - Y'_{601}\right)\right)
\end{aligned}
$$

To compute $Y'C_BC_R$ from $R'G'B'$ in the range [0...+1], scale the rows of the matrix of Eq 9.3 by the factors 219, 224, and 224, corresponding to the excursions of each of the components:

Eq 9.6

$$
\begin{bmatrix} Y'_{601} \\ C_B \\ C_R \end{bmatrix} = \begin{bmatrix} 16 \\ 128 \\ 128 \end{bmatrix} + \begin{bmatrix} 65.481 & 128.553 & 24.966 \\ -37.797 & -74.203 & 112. \\ 112. & -93.786 & -18.214 \end{bmatrix} \bullet \begin{bmatrix} R' \\ G' \\ B' \end{bmatrix}
$$

Summing the first row of the matrix yields 219, the luma excursion from black to white. The two entries of 112 reflect the positive C_BC_R extrema of the blue and red primaries.

To recover $R'G'B'$ in the range [0…+1] from $Y'C_BC_R$, use the inverse of Eq 9.6:

Eq 9.7

$$\begin{bmatrix} R' \\ G' \\ B' \end{bmatrix} = \begin{bmatrix} 0.00456621 & 0. & 0.00625893 \\ 0.00456621 & -0.00153632 & -0.00318811 \\ 0.00456621 & 0.00791071 & 0. \end{bmatrix} \bullet \left(\begin{bmatrix} Y'_{601} \\ C_B \\ C_R \end{bmatrix} - \begin{bmatrix} 16 \\ 128 \\ 128 \end{bmatrix} \right)$$

This looks overwhelming, but the $Y'C_BC_R$ components are integers in 8 bits, and the reconstructed $R'G'B'$ are scaled down to the range [0…+1].

Studio $R'G'B'$ signals often use the same 219 excursion as the luma component of $Y'C_BC_R$. To encode $Y'C_BC_R$ from $R'G'B'$ in the range [0…219], using 8-bit binary arithmetic, scale the encoding matrix of Eq 9.6 by $^{256}/_{219}$. Here is the encoding transform:

Eq 9.8

$$\begin{bmatrix} Y'_{601} \\ C_B \\ C_R \end{bmatrix} = \begin{bmatrix} 16 \\ 128 \\ 128 \end{bmatrix} + \frac{1}{256} \begin{bmatrix} 76.544 & 150.272 & 29.184 \\ -44.182 & -86.740 & 130.922 \\ 130.922 & -109.631 & -21.291 \end{bmatrix} \bullet \begin{bmatrix} R'_{219} \\ G'_{219} \\ B'_{219} \end{bmatrix}$$

You can determine the excursion that an encoding matrix is designed to produce, often 1, 219, 255, or 256, by summing the coefficients in the top row. In Eq 9.8, the sum is 256. If you find an unexpected sum, suspect an error in the matrix.

The multiplication by $^{1}/_{256}$ can be accomplished by shifting. For implementation in binary arithmetic the matrix coefficients have to be rounded. When you round, take care to preserve the row sums of [1, 0, 0]. After you add the offsets, clamp all three components to the range 1 through 254, inclusive, since Rec. 601 reserves codes 0 and 255 for synchronization signals.

To decode $R'G'B'$ in the range [0…219] from $Y'C_BC_R$, using 8-bit binary arithmetic:

Eq 9.9

$$\begin{bmatrix} R'_{219} \\ G'_{219} \\ B'_{219} \end{bmatrix} = \frac{1}{256} \begin{bmatrix} 256. & 0. & 350.901 \\ 256. & -86.132 & -178.738 \\ 256. & 443.506 & 0. \end{bmatrix} \bullet \left(\begin{bmatrix} Y'_{601} \\ C_B \\ C_R \end{bmatrix} - \begin{bmatrix} 16 \\ 128 \\ 128 \end{bmatrix} \right)$$

The entries of 256 in this matrix indicate that the corresponding component can simply be added; there is no need for a multiplication operation. This matrix contains entries larger than 256; the corresponding multipliers will need capability for 9 bits.

$Y'C_BC_R$ coding according to Rec. 601 is employed by component digital video equipment for 525/59.94 and 625/50, including D-1 digital videotape recorders. The digital representation of $Y'C_BC_R$ for HDTV is not yet standardized in Rec. 709.

The matrices in this section conform to Rec. 601 and apply directly to conventional 525/59.94 and 625/50 video. It is not yet decided whether emerging HDTV standards will use the same matrices, or adopt a new set of matrices having different luma coefficients. In my view it would be unfortunate if different matrices were adopted, because then image coding and decoding would depend on whether the picture was small (conventional video) or large (HDTV).

$Y'C_BC_R$ from computer RGB

In computing it is conventional to use 8-bit $R'G'B'$ components, with no headroom or footroom: Black is at code 0 and white is at 255. To encode $Y'C_BC_R$ from $R'G'B'$ in the range [0...255], using 8-bit binary arithmetic, the matrix of Eq 9.6 is scaled by $^{256}/_{255}$:

Eq 9.10

$$\begin{bmatrix} Y'_{601} \\ C_B \\ C_R \end{bmatrix} = \begin{bmatrix} 16 \\ 128 \\ 128 \end{bmatrix} + \frac{1}{256} \begin{bmatrix} 65.738 & 129.057 & 25.064 \\ -37.945 & -74.494 & 112.439 \\ 112.439 & -94.154 & -18.285 \end{bmatrix} \bullet \begin{bmatrix} R'_{255} \\ G'_{255} \\ B'_{255} \end{bmatrix}$$

This transform assumes that the $R'G'B'$ components have been subject to gamma correction; see page 107.

To decode $R'G'B'$ in the range [0...255] from Rec. 601 $Y'C_BC_R$, using 8-bit binary arithmetic:

Eq 9.11

$$\begin{bmatrix} R'_{255} \\ G'_{255} \\ B'_{255} \end{bmatrix} = \frac{1}{256} \begin{bmatrix} 298.082 & 0. & 408.583 \\ 298.082 & -100.291 & -208.120 \\ 298.082 & 516.411 & 0. \end{bmatrix} \bullet \left(\begin{bmatrix} Y'_{601} \\ C_B \\ C_R \end{bmatrix} - \begin{bmatrix} 16 \\ 128 \\ 128 \end{bmatrix} \right)$$

Some of these coefficients, when scaled by $^{1}/_{256}$, are larger than unity: 9-bit multipliers are required.

The matrix in Equation 9.11 will decode standard $Y'C_BC_R$ components to $R'G'B'$ components in the range [0...255], subject to roundoff error. You must take care to avoid overflow due to roundoff error. But you must protect against overflow in any case, because studio video signals use the extremes of the coding range to handle signal overshoot and undershoot, and these will require clipping when decoded to an RGB range that has no headroom or footroom.

Kodak PhotoYCC

The PhotoYCC color coding system was designed by Kodak for its PhotoCD system. Although designed for still pictures, not moving pictures, the system is optimized to display images on video equipment. The compression methods of the PhotoCD system are proprietary, and compression and decompression usually include conversion to and from RGB, so you are unlikely to encounter any raw image data in PhotoYCC form. However, PhotoCD has become a significant aspect of publishing, computing, and multimedia markets, so it deserves mention here.

PhotoYCC is broadly similar to $Y'C_BC_R$. PhotoYCC uses the Rec. 709 primaries, white point, and transfer function. Luma is coded with no footroom and lots of headroom, in order to accommodate highlights. Reference white corresponds approximately to luma code 189. The scaling of $B'-Y'$ and $R'-Y'$ is asymmetrical, with excursions quite different from those of Rec. 601, in order to accommodate a wide color gamut, similar to that of photographic film:

Eq 9.12

$$\begin{aligned} Y'_{601,189} &= \frac{255}{1.402} Y'_{601} \\ C_{1,8b} &= 156 + 111.40\left(B' - Y'\right) \\ C_{2,8b} &= 137 + 135.64\left(R' - Y'\right) \end{aligned}$$

For use with PhotoCD compression, the C_1 and C_2 components are subsampled by factors of 2 both horizontally and vertically. The subsampling should be

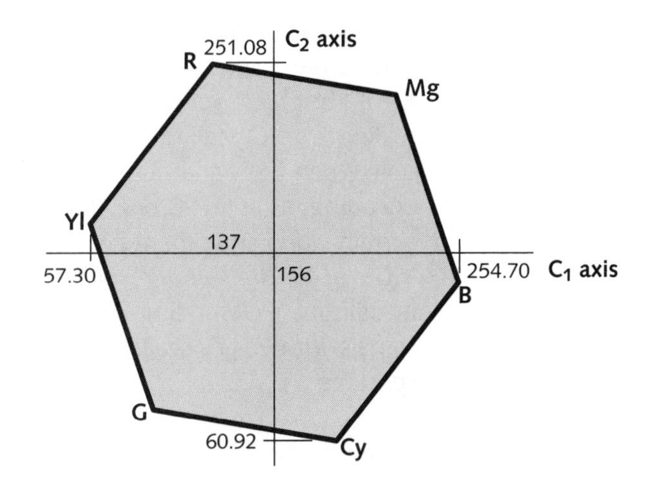

Figure 9.4 **PhotoYCC C_1C_2 components.**

considered a feature of the compression process and not of the PhotoYCC color coding.

The following equation is derived from Eq 9.12; it is comparable to Eq 9.7 in that it produces $R'G'B'$ in the range [0…+1] from integer YCC components:

Eq 9.13

$$\begin{bmatrix} R'_{709} \\ G'_{709} \\ B'_{709} \end{bmatrix} = \begin{bmatrix} 0.00549804 & 0. & 0.0051681 \\ 0.00549804 & -0.0015446 & -0.0026325 \\ 0.00549804 & 0.0079533 & 0. \end{bmatrix} \bullet \left(\begin{bmatrix} Y'_{601,189} \\ C_1 \\ C_2 \end{bmatrix} - \begin{bmatrix} 0 \\ 156 \\ 137 \end{bmatrix} \right)$$

Y'UV, Y'IQ confusion

I have detailed several coding systems based on $B'-Y'$ and $R'-Y'$ components: $Y'P_BP_R$, $Y'C_BC_R$, and PhotoYCC. These systems are all similar, but have different scale factors appropriate for different applications. In studio video, and in other applications that accommodate accurate color reproduction, the nomenclature for the scaling is quite precise. But outside these realms, the term *YUV* is applied loosely to any form of color difference coding based on $B'-Y'$ and $R'-Y'$ components. When the term *YUV* is encountered in connection with component digital video, generally $Y'C_BC_R$ is meant, but this is by no means guaranteed.

The following sections will describe (U, V) and (I, Q) color differences, which are also based on $B'-Y'$ and $R'-Y'$, but have yet another set of scale factors. UV scaling – or IQ scaling and rotation – is appropriate only when the signals are destined for composite encoding, as in NTSC or PAL. Do not be misled by certain video equipment having connectors labeled $Y'UV$ or Y', $B'-Y'$, $R'-Y'$, or by JPEG being described as utilizing $Y'UV$ coding. In fact the connectors convey signals with $Y'P_BP_R$ scaling, and the JPEG standard itself specifies $Y'C_BC_R$.

When the term YIQ is encountered in a computer graphics or image processing context, generally no mention is made of the transfer function of the underlying $R'G'B'$ components, and no account is taken of the nonlinear formation of luma. Image data supposedly coded in $Y'IQ$ is therefore suspect, especially since no analog or digital interface has been standardized for $Y'IQ$ components. Finally, any image data supposedly coded to the original 1953 NTSC primaries is also suspect, because it has been two decades since any equipment using these primaries has been built.

UV components

In the formation of NTSC, PAL, or S-video, it is necessary to scale the color difference components so that the eventual composite signal will be contained within the amplitude limits of the signal processing and recording equipment. It is standard to limit the composite excursion to the range -33 ⅓ to +133 ⅓ IRE units. To this end, the $B'-Y'$ and $R'-Y'$ signals are each scaled to form U and V components that satisfy this constraint:

Eq 9.14

$$-\frac{1}{3} \leq Y'_{601} \pm \sqrt{U^2 + V^2} \leq \frac{4}{3}$$

The scale factors are derived from two simultaneous equations involving $B'-Y'$ and $R'-Y'$, because the maximum excursions – at the blue and red primaries – are not located on the axes where scaling takes place.

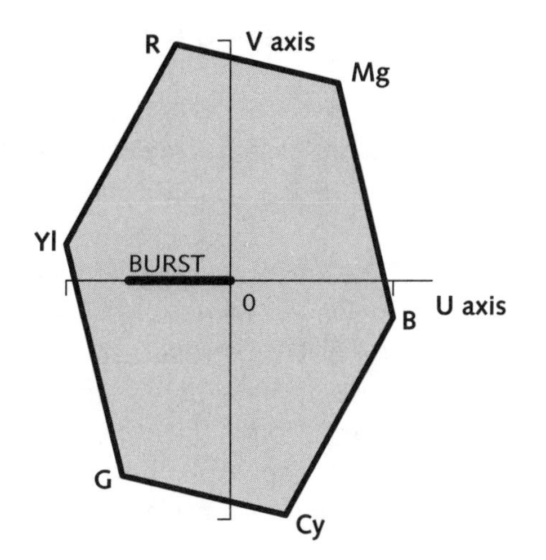

Figure 9.5 **UV components.**

$B'-Y'$ and $R'-Y'$ components are transformed to UV using these scale factors:

Eq 9.15

$$U = 0.492111\left(B' - Y'_{601}\right)$$
$$V = 0.877283\left(R' - Y'_{601}\right)$$

The scale factors are usually written rounded to three digits, because they were standardized in an era when that was the maximum precision that could be achieved. In matrix form the scaling is this:

Eq 9.16

$$\begin{bmatrix} Y'_{601} \\ U \\ V \end{bmatrix} = \begin{bmatrix} 1 & 0 & 0 \\ 0 & 0.492111 & 0 \\ 0 & 0 & 0.877283 \end{bmatrix} \bullet \begin{bmatrix} Y'_{601} \\ B' - Y'_{601} \\ R' - Y'_{601} \end{bmatrix}$$

To obtain $Y'UV$ from $R'G'B'$, concatenate the matrix above with the matrix of Equation 9.1 on page 172:

Eq 9.17

$$\begin{bmatrix} Y'_{601} \\ U \\ V \end{bmatrix} = \begin{bmatrix} 0.299 & 0.587 & 0.114 \\ -0.147407 & -0.289391 & 0.436798 \\ 0.614777 & -0.514799 & -0.099978 \end{bmatrix} \bullet \begin{bmatrix} R' \\ G' \\ B' \end{bmatrix}$$

To recover $R'G'B'$ from $Y'UV$, invert that matrix:

Eq 9.18

$$\begin{bmatrix} R' \\ G' \\ B' \end{bmatrix} = \begin{bmatrix} 1 & 0 & 1.140251 \\ 1 & -0.393931 & -0.580809 \\ 1 & 2.028398 & 0 \end{bmatrix} \bullet \begin{bmatrix} Y'_{601} \\ U \\ V \end{bmatrix}$$

The $+\frac{4}{3}$ limit applies to composite studio equipment; however, a limit of +1.2 applies to terrestrial (VHF/UHF) broadcast. Many computer graphics systems have provisions to check or limit chroma excursion accordingly. A practical limit of approximately +1.1 is appropriate to avoid interference problems with sound in older television receivers.

Y'UV coding is unique to NTSC, PAL and SECAM: It is not used in component video, is not used in HDTV, and should not be used in computing.

IQ components

Vision has less spatial acuity for purple-green transitions than it does for orange-cyan. The *U* and *V* signals of *Y'UV* must be carried with equal bandwidth, albeit less than that of luma, because neither aligns with the minimum color acuity axis. However, if signals *I* and *Q* are formed from *U* and *V* as I will describe, then the *Q* signal can be more severely filtered than *I* – to about 600 kHz, compared to about 1.3 MHz – without loss in chroma resolution being perceptible to a viewer at typical television viewing distance. The *IQ* pair is equivalent to *UV* with a 33° rotation and an exchange of axes.

Certain texts have transformation matrices that erroneously omit the exchange of axes.

The rotation and exchange can be accomplished by this matrix transform:

Eq 9.19

$$\begin{bmatrix} Y'_{601} \\ I \\ Q \end{bmatrix} = \begin{bmatrix} 1 & 0 & 0 \\ 0 & -0.544639 & 0.838671 \\ 0 & 0.838671 & 0.544639 \end{bmatrix} \bullet \begin{bmatrix} Y'_{601} \\ U \\ V \end{bmatrix}$$

The matrix is its own inverse, so the same matrix recovers *Y'UV* from *Y'IQ*.

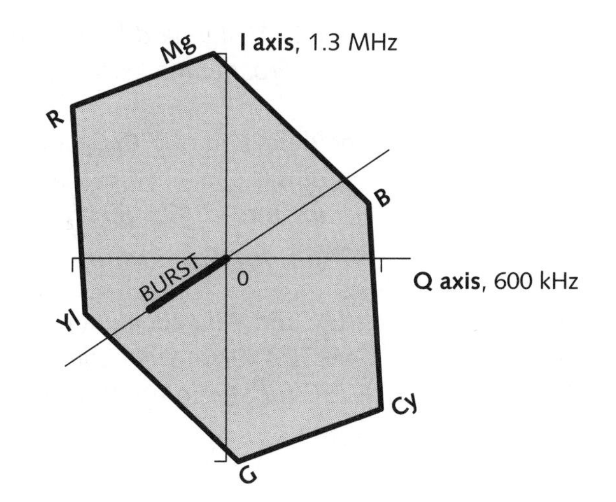

Figure 9.6 **IQ components.**

To obtain the encoding from $R'G'B'$, the rotation and exchange above are concatenated with Eq 9.17:

Eq 9.20

$$\begin{bmatrix} Y'_{601} \\ I \\ Q \end{bmatrix} = \begin{bmatrix} 0.299 & 0.587 & 0.114 \\ 0.595879 & -0.274133 & -0.321746 \\ 0.211205 & -0.523083 & 0.311878 \end{bmatrix} \bullet \begin{bmatrix} R' \\ G' \\ B' \end{bmatrix}$$

To recover $R'G'B'$ from $Y'IQ$, use this relation:

Eq 9.21

$$\begin{bmatrix} R' \\ G' \\ B' \end{bmatrix} = \begin{bmatrix} 1 & 0.956295 & 0.621025 \\ 1 & -0.272558 & -0.646709 \\ 1 & -1.104744 & 1.701157 \end{bmatrix} \bullet \begin{bmatrix} Y'_{601} \\ I \\ Q \end{bmatrix}$$

$Y'IQ$ coding is unique to NTSC: PAL and SECAM systems use equiband U and V components.

Color coding standards

The definition of $Y'C_BC_R$ for component digital video in 525/59.94 and 625/50 systems is contained in Recommendation ITU-R (formerly CCIR) BT.601-4, *Encoding Parameters of Digital Television for Studios* (Geneva: ITU, 1990).

$Y'C_BC_R$ for component digital HDTV, and $Y'P_BP_R$ for component analog HDTV, are standardized in Recommendation ITU-R BT.709, *Basic Parameter Values for the HDTV Standard for the Studio and for International Programme Exchange* (Geneva: ITU, 1990).

The definition of $Y'C_BC_R$ and $Y'P_BP_R$ for 1125/60 systems is contained in SMPTE 240M-1992, *Signal parameters – 1125/60 high-definition production system*.

Y'UV and *Y'IQ* coding are described in SMPTE 170M-1994, *Composite Analog Video Signal – NTSC for Studio Applications*.

Composite NTSC and PAL 10

Video can use *R'G'B'* components directly, but three signals are expensive to record, process, or transmit. Providing some loss of color detail is acceptable, then luma *Y'* and color difference components based on *B'–Y'* and *R'–Y'* can be used, as explained in *Component video color coding*, on page 171. But even that method has a fairly high information rate (bandwidth, or data rate).

Composite coding uses *quadrature modulation* to combine the two color difference components into a *chroma* signal, and *frequency interleaving* to combine luma and chroma into one signal of less bandwidth or data rate.

In principle, NTSC or PAL coding could be used with any scanning standard. But in practice NTSC and PAL are used only with 525/59.94 and 625/50 scanning, and the parameters of encoding are optimized for those scan frequencies. This chapter presents the concepts of composite encoding; for more detail see *525/59.94 NTSC composite video*, on page 221, and *625/50 PAL composite video*, on page 241.

Composite encoding was invented to limit transmitted bandwidth at the introduction of color television, and it has proven highly effective for broadcast. In the studio, composite encoding is no longer necessary: The bandwidth to carry color video in component form is now generally affordable. In the composite domain it is impossible to perform many processing operations: Even to resize a picture requires decoding, processing, and reencoding. The difficulty of performing image manipulation on composite signals is an incentive to use component video systems. But there is a huge installed base of composite equipment, and there remains a cost premium associated with component equipment.

The frequency interleaving inherent in NTSC and PAL introduces a certain degree of mutual interference between luma and chroma. Once a signal has been encoded to composite form, it suffers from the NTSC or PAL *footprint:* Some degree of interference between luma and chroma has irreversibly been impressed on the signal, no matter what subsequent steps are taken.

The S-video interface was first introduced in S-VHS VCRs, and S-video remains a feature of all S-VHS VCRs. But the S-video interface is quite independent of S-VHS recording technology.

Although *Y'/C* (S-video) employs quadrature modulation, it does not use frequency interleaving, so S-video avoids the footprint. However, S-video has limited chroma bandwidth compared to 4:2:2.

Subcarrier regeneration

Chroma demodulation depends on the decoder having access to the continuous-wave subcarrier used in encoding. To this end, an encoder inserts a brief sample of the inverted cos subcarrier, *burst*, into the horizontal sync interval. A decoder regenerates subcarrier using a circuit whose block diagram is shown in Figure 10.1 below: Continuous-wave *cos* and *sin* are generated in a crystal oscillator whose frequency and phase are updated once per line by a phase comparator, based on a comparison of the signal's burst.

Figure 10.1 **Subcarrier regeneration.**

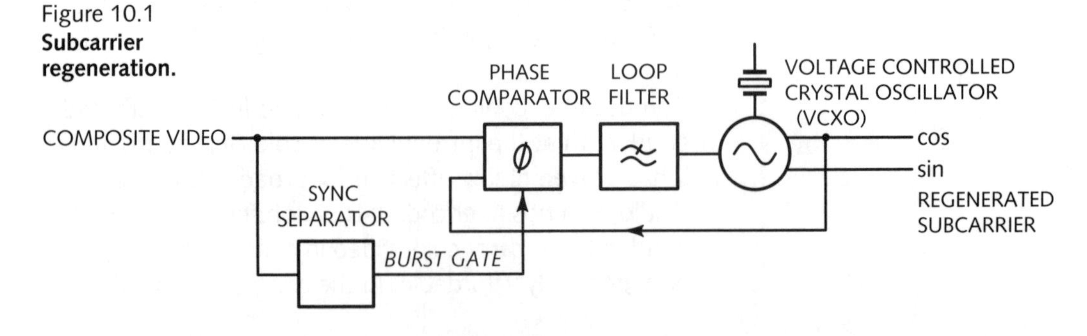

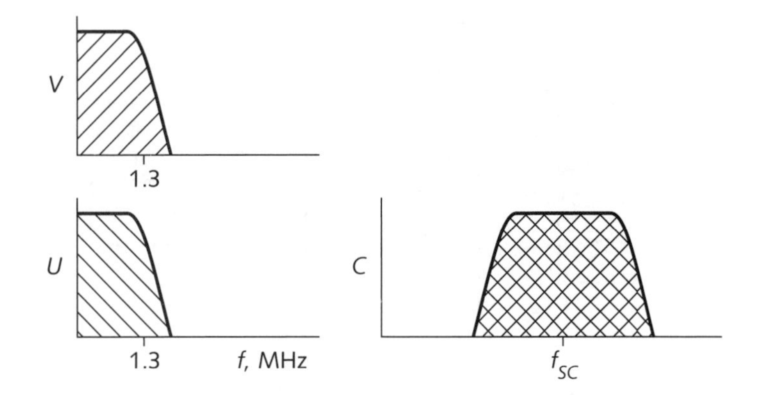

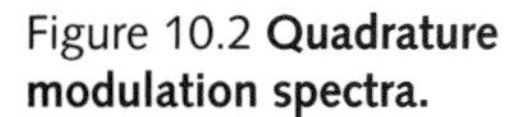
Figure 10.2 **Quadrature modulation spectra.**

Quadrature modulation

The process of NTSC encoding starts with luma *Y′* and two color difference components, *U* and *V*, that have been scaled from *B′–Y′* and *R′–Y′* as described in *UV components*, on page 180. The scaling limits the excursion of the eventual composite signal. The *U* and *V* color difference components are subjected to lowpass filtering to about 1.3 MHz; the resulting spectra are sketched at the left of Figure 10.2 above.

The *U* and *V* components are combined into a single *chroma* signal *C* using *quadrature modulation*, which involves multiplying *U* and *V* by *cos* and *sin* of the subcarrier, then summing and filtering the products:

Eq 10.1

$$C = U(\cos t) + V(\sin t)$$

In this equation, cos *t* represents the subcarrier, typically about 3.58 MHz or 4.43 MHz. The exact choice of subcarrier frequency is not particularly important to quadrature modulation, but it is very important to frequency interleaving, as explained in *Frequency spectrum of NTSC*, on page 69. The spectrum of the modulated chroma is sketched at the right of Figure 10.2. Chroma is centered on the subcarrier frequency; it has a lower sideband extending about 1.3 MHz below the subcarrier frequency, and an upper sideband extending about 1.3 MHz above.

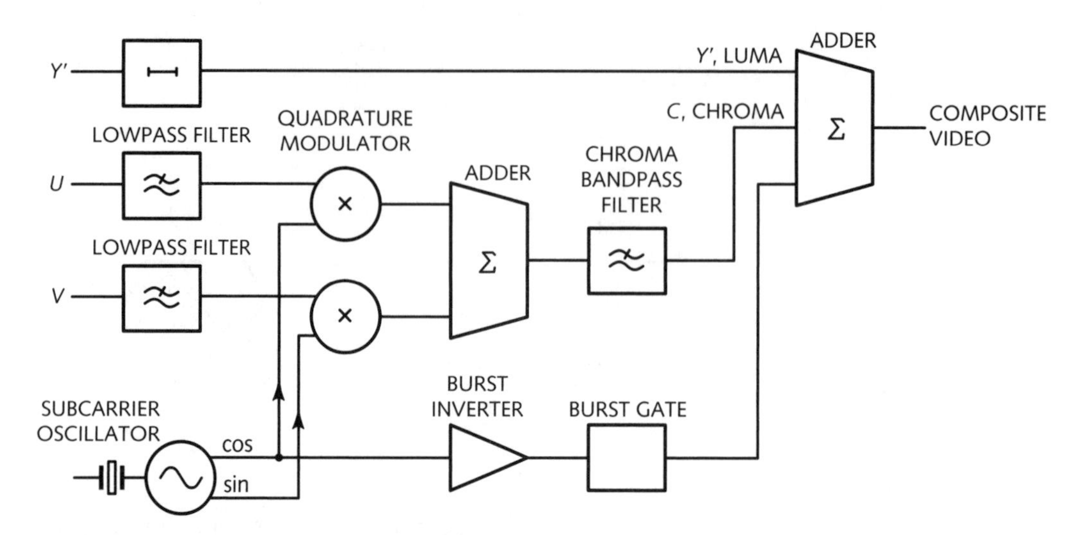

Figure 10.3 **NTSC encoder block diagram.** A subcarrier oscillator generates *cos* and *sin* continuous waves. Quadrature modulation is performed on lowpass-filtered *U* and *V* components by a pair of 4-quadrant multipliers. An adder forms composite chroma, *C*. Chroma is summed with luma *Y'* to form composite video. Frequency interleaving is achieved when subcarrier and scan rates are coherent. In studio video, subcarrier is usually divided from the same master clock that generates the raster.

Once modulated chroma has been formed, composite video is formed by summing luma and chroma. Figure 10.3 above shows the block diagram of an NTSC encoder. An S-video interface uses the *Y'* and *C* signals after modulation but prior to summation.

An NTSC decoder is shown in Figure 10.4 opposite. Decoding begins with separation of the luma and chroma components; details of *Y'/C separation* will be described in a moment. Separated chroma is then multiplied simultaneously by *cos* and *sin* of the regenerated subcarrier, and the products are lowpass filtered to recover the *U* and *V* components. If a decoder has an S-video interface, *Y'* and *C* signals are available at the interface, and the *Y'/C* separator is bypassed.

Quadrature modulation is reversible without information loss, provided that *U* and *V* are limited in bandwidth to less than half the subcarrier frequency. In practice, the modulation itself introduces no significant impairments, although the bandwidth limitation of the color difference signals removes color detail.

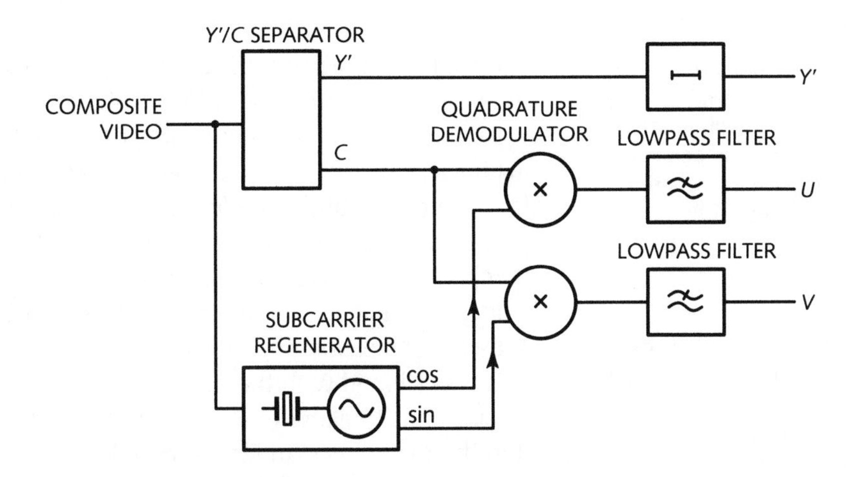

Figure 10.4 **NTSC decoder block diagram.** In a decoder, luma and chroma are separated by a notch filter or a comb filter. Subcarrier is regenerated from the burst. Continuous *sin* and *cos* waves in quadrature are presented to a pair of multipliers; the products are lowpass filtered to recover *U* and *V*.

Decoder and monitor controls

A few decades ago, consumer receivers had unstable circuitry: User adjustment was necessary to produce acceptable color. Modern circuits are so stable that user controls are no longer necessary, but consumers demand that they continue to be provided.

Monitor controls

Black level, sometimes misleadingly named *Brightness*, adjusts the offset of the red, green, and blue components presented to the display. *Picture*, sometimes misleadingly named *Contrast*, adjusts overall gain (and thereby adjusts the intensity of reproduced white). These two controls operate in the *R'G'B'* domain; they are unrelated to NTSC or PAL decoding. The user controls affect red, green, and blue identically; a CRT monitor has a set of internal calibration adjustments called *Screen* and *Drive* that affect the offset and gain of the individual red, green, and blue components.

Decoder controls

There are two main decoder controls. One control adjusts the phase of the regenerated subcarrier. Its effect is to rotate hue of the decoded colors; it is called *Hue* or *Tint*. The other control affects the gain of the chroma component, or equivalently, the decoded color difference components. It is called *Saturation* or *Color*.

Narrowband I chroma

Hazeltine Corporation, *Principles of Color Television*, by the Hazeltine Laboratories staff, compiled and edited by Knox McIlwain and Charles E. Dean. New York: John Wiley & Sons, 1956.

With subcarrier at about 3.6 MHz, modulated chroma extends to about 4.9 MHz. The designers of NTSC were faced with a channel bandwidth of 4.2 MHz, insufficient to convey the upper sideband of modulated chroma. Bandwidth limitation of the composite signal would cause cross-contamination of the chroma components. This led to a scheme whereby *I* and *Q* color difference components were formed from *U* and *V* by a 33° rotation and an exchange of axes.The *Q* axis was aligned with the minimum color acuity axis of vision, so as to allow *Q* to be more severely bandlimited than *I*. It was standard to bandlimit *I* to 1.3 MHz and *Q* to 0.6 MHz (600 kHz). Quadrature modulation was then performed with subcarrier phase-shifted 33°.

Some documents and texts mistakenly omit the exchange of *B'−Y'* and *R'−Y'* axes in describing *Y'IQ* coding. Other texts give the mistaken impression that *Y'IQ* is in use with linear-light coding.

At the decoder, a special filter could recover the low frequency component of *I*, from 0 to 0.6 MHz, from the lower and upper sidebands of modulated chroma, and recover the high frequency *I* component – from 0.6 MHz to 1.3 MHz – from the lower sideband alone.

The benefit of this aspect of NTSC's design was never realized, because consumer receivers rarely implemented recovery of the wideband *I* component. The bandwidth realizable in consumer equipment, especially VCRs, and the technical difficulty of encoding *Y'IQ* properly in the studio, had the effect of limiting the maximum usable color difference bandwidth to about 600 kHz in both components. The recently adopted SMPTE 170M standard effectively endorses narrowband *U* and *V* coding for broadcast.

SMPTE 170M encourages wideband *U* and *V* in the studio, but permits wideband *I* and narrowband *Q*. Chroma encoding with wideband *U* and *V* causes the composite signal to occupy a bandwidth of about 5.5 MHz; filtering to about 4.2 MHz is necessary prior to broadcast. Full *I* bandwidth could be retained if, prior to filtering, chroma was decoded, then reencoded using wideband *I* and narrowband *Q*. decoding and reencoding is virtually never undertaken in practice: Instead, the composite signal is just filtered, causing irreversible cross-contamination between the highband portions of *U* and *V*. The result is that wideband *U* and *V* are typically present in the studio, but only narrowband chroma is broadcast.

Once modulated chroma is formed, an analog NTSC decoder cannot determine whether the encoder operated with *Y'UV* or *Y'IQ* components: A demodulator cannot detect whether the encoder was operating at 0° or 33° phase. So in analog usage, the terms *Y'UV* and *Y'IQ* are often used somewhat interchangeably.

Composite $4f_{SC}$ digital video equipment, including D-2 VTRs, use NTSC encoded on the *I* and *Q* axes.

Frequency interleaving

As I explained in *Frequency spectrum of NTSC*, on page 69, NTSC subcarrier inverts phase line to line with respect to sync. In component video, every frame of a still image is represented identically, but in composite NTSC a still image is represented by two different composite *colorframes,* denoted *A* and *B*. The line-to-line inversion of subcarrier phase produces the frequency interleaving by which luma and chroma can be accurately separated using a comb filter.

The line-to-line inversion of subcarrier, combined with the odd number of lines in a frame required to produce interlace, causes subcarrier to invert frame to frame. This inversion is almost incidental, although it permits the use of a frame comb to separate luma and chroma.

On page 69, I explained how NTSC luma and chroma can be separated using notch filtering and comb filtering. A decoder with a notch filter will produce *cross-color* artifacts, that may appear as swirling rainbows, when luma occupying frequencies in the range of subcarrier is mistakenly decoded as chroma. A notch filter can also introduce *cross-luma* artifacts, when chroma is mistakenly decoded as luma.

In a television receiver that uses a notch filter, chroma inverts in phase line by line even if mistakenly decoded as luma. When integrated vertically by large spot size at the display, and when integrated vertically by the viewer's visual system, the visibility of cross-luma artifacts is minimized. This is the mechanism that allowed color to be retrofitted into monochrome NTSC: the newly added color component was not overly visible.

One cross-luma artifact deserves special mention. Consider a set of scan lines, each containing the same large, abrupt change in color – say from one colored vertical bar to another. The modulated chroma produced by this transition can extend far past the subcarrier frequency, while still being contained in the chroma region. Most decoders will mistakenly interpret

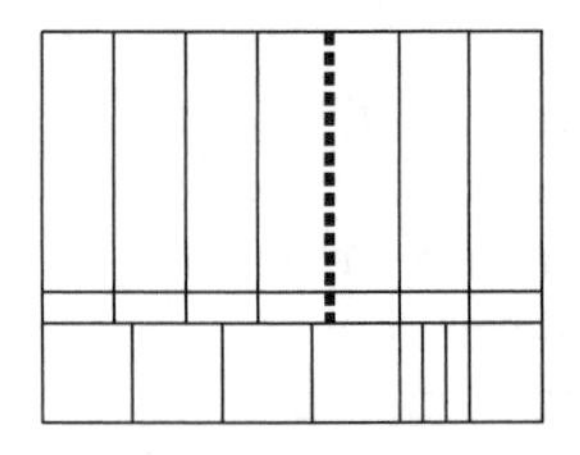

Figure 10.5 **Dot crawl.**

some of the power in this transition to be luma. The apparent luma will alternate in phase from line to line, and will also invert frame to frame. The frame-rate inversion, combined with interlace, produces a fine pattern of dots that apparently traverse upward along the transition at a rate of one scan line per field, about eight seconds through the height of the image. This version of cross-luma is called *dot crawl*. It is avoided entirely by the use of a comb filter at the decoder.

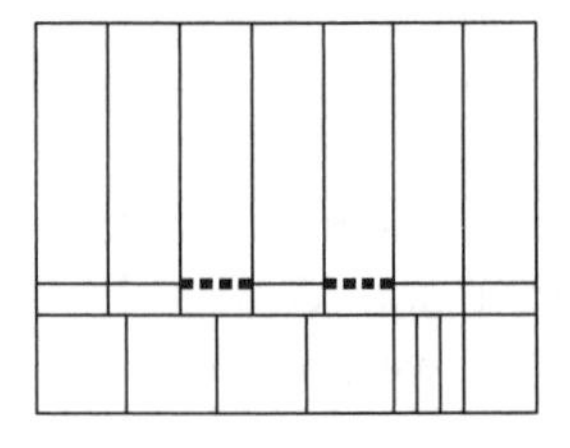

Figure 10.6 **Hanging dots.**

A particular cross-color artifact is apparent on the ubiquitous colorbar test signal that will be described on page 257. The colorbar pattern involves an abrupt transition between a line containing one set of saturated colors, and a line containing a completely different set of colors. When decoded by a comb filter, the abrupt vertical transition contains power that decodes to a strong luma component at the subcarrier frequency. On a monitor with sufficiently high resolution, a stationary, horizontal line of *hanging dots* is displayed at the transition. The artifact is strikingly obvious when colorbars are displayed on a studio monitor equipped with a comb filter. But the effect is seldom seen on consumer receivers, and seldom seen in actual pictures.

Yves Faroudja's SuperNTSC system includes comb filter preprocessing at the encoder. This prevents the worst of cross-color and cross-luma artifacts, even with notch filter decoders. However, SuperNTSC encoders are not widely deployed.

The introduction of cross-luma and cross-chroma artifacts at a decoder can be minimized by using a comb filter. However, it is uncommon for an *encoder* to perform any processing on luma and chroma prior to their being summed. If luma and chroma components overlap in the spatial frequency domain, luma and chroma will be confused at the encoder, and no subsequent steps at a decoder can repair the damage. Upon summation of Y' and C signals at an encoder, the composite *footprint* is said to be imposed on the signal.

PAL encoding

PAL encoding involves quadrature modulation of U and V color difference components having equal 1.3 MHz bandwidths. In PAL, the subcarrier is at 4.43 MHz, and the composite signal bandwidth is about 6 MHz. Equiband U and V color difference components are used: Because video bandwidth in PAL systems is sufficient to accommodate the entire 1.3 MHz of modulated chroma, PAL has no need for *I* and *Q* components.

In NTSC, the subcarrier inverts phase on alternate lines. The *A* in PAL refers to the alternation not of subcarrier, but of the V chroma component.

In a PAL encoder, the NTSC block diagram is augmented by a *V-axis switch*, which alternates the phase of the modulated *V* component on a line by line basis. PAL derives its acronym from this *Phase Alternation at Line rate*. Associated with the V-axis switch, burst alternates at line rate, between +135° and -135°, with respect to subcarrier (compared to a fixed 180° burst phase for NTSC).

PAL differs from NTSC in several other details. In studio NTSC, subcarrier frequency is an odd multiple of twice the line rate. In PAL, subcarrier frequency is based on an odd multiple of one-quarter the line rate, leading to roughly a 90° advance of subcarrier phase line by line. This phase advance, combined with the odd number of lines in a frame, leads to PAL having four *colorframes* denoted *I, II, III,* and *IV.*

In standard B/PAL, G/PAL, H/PAL, and I/PAL, a *+25 Hz offset* is added to this basic subcarrier frequency to reduce the visibility of an artifact called *Hannover bars*. The +25 Hz frequency offset contributes an additional 0.576° to the line by line advance of subcarrier phase. Since the offset adds one complete subcarrier cycle per frame, it has no impact on the four-color-frame sequence.

PAL color coding has the advantage over NTSC that it is inherently immune to hue shifts – to which the eye is quite sensitive – caused by a transmission impairment called *differential phase* error. This advantage is offset by increased complexity both at the transmitter and at the receiver: A one-line (1*H*) delay is required at the receiver to take full advantage of this property; most PAL receivers benefit from the resulting comb filter.

While PAL's modifications to NTSC once conferred an incremental performance advantage, that has long since been subverted by the great difficulty of exchanging program material between countries, and by the expense of producing professional and consumer equipment that operates with both standards. These features of PAL must now be seen as premature optimization. I hope that European television engineers will think twice before again adopting a uniquely European standard. But history stands ready to repeat itself: Europe is poised to adopt a unique HDTV scanning standard, 1250/50.

PAL-M, PAL-N

Conventional 625/50 PAL is broadcast using ITU-R systems *B*, *G*, *H*, and *I*. In South America, PAL variants PAL-M and PAL-N are used. You can find details in *M/PAL (PAL-M, PAL-525)* and *M/PAL (PAL-M, PAL-525)*, on page 253. The gratuitous parameter differences introduced by these systems have no technical merit; they have served the purpose of deterring image exchange and equipment commonality. Today there is no production equipment for these standards: Production is accomplished using standard 525/59.94 or 625/50 equipment, according to the appropriate raster standard, and the signal is transcoded to PAL-M or PAL-N immediately prior to transmission.

Incoherent subcarrier

In studio-quality NTSC video, the color subcarrier has a frequency of exactly 227.5 times the line rate. Although its phase may be subject to some uncertainty, its frequency is *coherent* with line rate.

When subcarrier and sync are generated in a single circuit, it is simple to arrange for coherence of the subcarrier and the line rate: The two need simply be divided from the same master oscillator. But low-cost consumer and computer systems have no provision to generate coherent subcarrier, and in any case the tolerance on line rate would be an order of magnitude worse than that required for broadcast.

If a decoder is presented with an input signal whose subcarrier is incoherent with its sync, the frequency-interleaving principle should be abandoned: Chroma and luma should be separated by lowpass and bandpass filters, not a comb filter. In this case the decoder might as well limit luma bandwidth to about 3 MHz, and chroma bandwidth to about 500 kHz; components outside this range will suffer serious artifacts.

Analog videotape recording

Analog videotape tape recorders (VTRs) and videocassette recorders (VCRs) involve mechanical scanning of the tape using a rotating tape head. Instabilities in the mechanical scanning during recording and playback introduce timebase error into the video signal. This timebase error can be removed by a subsystem or piece of equipment called a *timebase corrector.* A TBC recovers timing information – a *jittered clock* – from the sync of the off-tape signal. The TBC writes the unstable signal into a buffer memory using the jittered clock, and reads it out several line times later based on a crystal-stable clock. TBCs were once stand-alone units; they are now usually internal to broadcast VTRs.

There are three types of analog VTRs: *component, direct color,* and *color under* (or *heterodyne*).

Component VTR

A component VTR records component video in three channels. The recorded signal has no subcarrier, and although timebase correction is usually necessary, no subcarrier-related processing is needed upon playback.

Direct-color VTR

High-quality composite analog VTRs (such as 1-inch Type C) frequency-modulate the video signal onto an RF carrier, and record the modulated composite signal on tape. A direct-color VTR maintains the coherence of subcarrier to line rate, while introducing long-term (field-to-field) timebase jitter of perhaps ±5 video line times, and short-term timebase error (line-to-line) of up to perhaps ±500 ns. The short-term frequency variation makes phase-locking a color subcarrier crystal impossible upon playback. A direct-color videotape playback signal must be processed through a TBC before viewing or further processing.

Heterodyne VTR

Heterodyne (or *color under*) video recording is ubiquitous in low-cost videocassette recorders, including 3⁄4-inch (U-matic), 1⁄2-inch Betamax, VHS, 8 mm (Video-8), and Hi-8. The luma and chroma components of the input signal are separated prior to recording. Luma is recorded by FM in a manner similar to direct recording. Chroma is mixed down to a low-frequency subcarrier between 600 kHz and 1 MHz; upon recording, this signal penetrates deeply into the tape medium. In the formation of the low-frequency color subcarrier, the coherence of subcarrier and line rate is lost.

When color under recording was introduced, in U-matic VCRs, the low frequency subcarrier was approximately 688 kHz. A *dub connector* carried the separate components; this interface was known as *Y′C688*. The original VHS VCR used a low-frequency subcarrier at about 629 kHz; some industrial equipment implements the *Y′C629* interface.

Upon playback, the color-under chroma signal is remodulated onto a crystal-stable color subcarrier and summed with luma. In most color-under VCRs, luma bandwidth is less than 2.5 MHz, and chroma bandwidth is less than 500 kHz.

Although playback chroma from a color under VCR is remodulated onto a crystal-stable subcarrier, the luma and chroma signals – and the sync component of both – contain timebase errors. But the horizontal oscillator of a television receiver has sufficient range to track the timebase error, and a crystal oscillator can lock to the subcarrier. So a TBC is not necessary upon playback to a monitor, though some residual timebase jitter will be visible if one is not used.

After color-under recording, the frequency interleaving that may have once been present in the signal is lost from the composite interface. There is no point in attempting to extract more than about 3 MHz of luma bandwidth. However, if an S-video interface is implemented in a color under VCR, the signals have been processed by the recorder without separation or combining of luma and chroma. Much better performance can be obtained from the S-video interface than from the composite interface.

NTSC-4.43

Some multistandard consumer equipment implements a system known as *NTSC-4.43*, whereby 525/59.94 NTSC tapes are played back using the subcarrier frequency and chroma modulation of PAL instead of the subcarrier frequency and chroma modulation of NTSC. A 625/50 receiver having a vertical lock range sufficient to accommodate the 60 Hz field rate can recognize the special mode, and demodulate chroma using its 4.43 MHz color subcarrier crystal. NTSC-4.43 is a feature of a VCR, a monitor, or a receiver. It is not a tape format and is not used for broadcast.

SECAM

The SECAM system is used for terrestrial VHF and UHF transmission in France, Russia, and a few other countries. SECAM uses neither quadrature modulation nor frequency interleaving. SECAM has luma identical to PAL, but luma is summed with line-alternate *U* and *V* color difference components that are frequency-modulated (FM) onto a carrier roughly in the frequency range of PAL or NTSC color subcarrier. Since the color difference components are transmitted line-sequentially, a SECAM receiver requires a one-line (*1H*) delay to recover chroma. SECAM does not have a subcarrier *per se*, but has a pair of reference *rest* frequencies that are locked to the line rate. Each video line in SECAM includes a burst at the rest frequency for the line.

SECAM had an advantage during the 1960s and 1970s that because color information is recorded using FM, the color information of a signal with timebase error could be recovered without a timebase corrector. However, FM chroma encoding makes a SECAM signal nonlinear: A SECAM signal cannot be processed in composite form, even for an operation as simple as fading between two signals. SECAM is used today only as a transmission standard: No contemporary production equipment uses SECAM. Production for SECAM is accomplished using 625/50 component or composite PAL equipment, and the signal is transcoded to SECAM immediately prior to transmission.

Field, frame, line, and sample rates 11

This chapter outlines the field, frame, line, and sampling rates of 525/59.94 and 625/50 video. The standard sampling frequency for component digital video is exactly 13.5 MHz; this rate produces an integer number of samples per line in both 525/59.94 and 625/50. Modern systems such as HTDV sample at multiples of this rate.

Field rate

Television systems originated with field rates based on the local AC power line frequency: 60 Hz for North America and 50 Hz for Europe. The reason for this choice was that coupling of the ripple of a receiver's power-supply into circuitry – such as video amplifiers and high-voltage supplies – had an effect on the instantaneous brightness of the display. Interference caused artifacts called *hum bars*. The visibility of hum bars was minimized by making them stationary; this was done by choosing a field rate the same as the power line frequency. There was no requirement to have a exact frequency match, or to lock the phase: As long as the pattern was stationary, or drifting very slowly, it was not objectionable. The power supply interactions that were once responsible for hum bars no longer exist in modern circuitry, but the vertical scan rates that were standardized remain with us.

Line rate

The number of lines chosen for each system is the product of a few small integers: 525 is $7 \times 5^2 \times 3$, and 625 is 5^4. The use of small integer factors arose from the use of vacuum tube divider circuits to derive the field rate from the line rate: These dividers were stable only for small division ratios.

In both 525-line and 625-line systems, the total number of scan lines per frame is odd. Equivalently, the field rate is an odd multiple of half the line rate. This relationship generates the 2-to-1 interlace that reduces the visibility of wide-area flicker for a given resolution and transmission bandwidth.

These factors combined to give monochrome 525/60 television a line rate of $60 \times (7 \times 5^2 \times 3)$, or exactly 15.750 kHz. Monochrome 625/50 television has a line rate of $5 \times 5 \times 5 \times 5$ or exactly 15.625 kHz.

Sound subcarrier

In about 1941, the first NTSC recognized that visibility of sound-related patterns in the picture could be minimized if the picture line rate and the sound subcarrier rest frequency were coherent. In monochrome 525/59.94 television the sound subcarrier was placed at 4.5 MHz, exactly $^{2000}/_{7}$ (or 285 $^{5}/_{7}$) times the line rate. Sound in conventional television is frequency-modulated, and with an analog sound modulator even perfect silence cannot be guaranteed to generate an FM carrier of exactly 4.5 MHz. Nonetheless, making the FM sound carrier average out to 4.5 MHz was thought to have some value.

Addition of composite color

With the introduction of color, the two systems chose the same basic frequency interleaving technique to fit compatible color into their existing monochrome systems. In this technique, the color subcarrier frequency is chosen to alternate phase line by line, so as to minimize the visibility of encoded color on a monochrome receiver. This line-to-line phase relationship makes it possible to accurately separate chroma from luma when the decoder incorporates a line delay (comb filter), although a cheaper notch filter can be used instead. For details, see *Frequency spectrum of NTSC*, on page 69.

NTSC color subcarrier

In 1953 the second NTSC determined that it would be appropriate to choose a color subcarrier in the region of 3.6 MHz. They recognized that any nonlinearity in the processing of the composite color signal with sound – such as limiting in the *intermediate frequency* (IF) stages of a receiver – would result in intermodulation distortion between the sound subcarrier and the color subcarrier. The difference, or *beat frequency,* between the two subcarriers – about 920 kHz – falls in the luminance bandwidth and could potentially have been quite visible.

The NTSC recognized that the visibility of this pattern could be reduced if the beat frequency was line-interlaced. Since the color subcarrier is necessarily an odd multiple of half the line rate, the sound subcarrier had to be made an integer multiple of the line rate.

The NTSC decided that the color subcarrier should be exactly $^{455}/_{2}$ times the line rate. Line interlace of the beat could be achieved by increasing the sound-to-line rate ratio, previously 285 $^{5}/_{7}$, by the fraction $^{1001}/_{1000}$ to the next integer 286.

Responsibility for setting broadcast standards resided with the Federal Communications Commission. If the FCC had allowed the sound subcarrier rest frequency to be increased by the fraction $^{1001}/_{1000}$ – that is, increased by 4.5 kHz to about 4.5045 MHz – then the color subcarrier in NTSC would have been exactly 3.583125 MHz, the original line and field rates would have been unchanged, we would have retained exactly 60 frames per second, and NTSC would have no dropframes! Since sound is frequency-modulated, the sound carrier was never crystal-stable at the subcarrier frequency anyway – not even during absolute silence – and the tolerance of the rest frequency was already reasonably large (±1 kHz). The deviation of the sound subcarrier was – and remains – 25 kHz, so a change of 4.5 kHz could easily have been accommodated by the intercarrier sound systems of the day.

$$525 \times \left(\frac{60}{2} \text{Hz} \right) \times \frac{1000}{1001} \times \frac{455}{2}$$
$$= \frac{315}{88} \text{ MHz}$$
$$\approx 3.579\dot{5}4\dot{5} \text{ MHz}$$

The FCC standardized the nominal subcarrier value at 3.579545000, even though the designers understood that the decimal fraction recurred.

But the FCC refused to alter the sound subcarrier. Instead, the color/sound constraint was met by reducing both the line rate and field rate by the fraction $^{1001}/_{1000}$, to about 15.734 kHz and 59.94 Hz. Color subcarrier became 3.579545^{+} MHz.

The factors of 1001 are 7, 11, and 13. This numerical relationship was known in ancient times: The book *1001 Arabian Nights* is based on it. The numbers 7, 11, and 13 are considered to be very unlucky. Unfortunately the field rate of $^{60}/_{1.001}$, about 59.94 Hz, means that 60 fields consume slightly more than one second: Counting 30 fields per second does not agree with clock time. Dropframe timecode was invented to alleviates this difficulty; see *Timecode*, on page 265.

$$525 \times \left(\frac{60}{2} \text{Hz} \right) \times \frac{1000}{1001} \times 455 \times 2$$
$$= \frac{315}{22} \text{ MHz}$$
$$\approx 14.318\dot{1}8\dot{1} \text{ MHz}$$

NTSC sync generators generally use a master oscillator of about 14.318 MHz. The master clock is divided by 4 to obtain color subcarrier, and divided by 7 to obtain a precursor of line rate. Ultrastable television systems have atomic clocks that provide 5 MHz, followed by a rate multiplier of $^{63}/_{22}$ to derive the master 14.318 MHz clock.

PAL color subcarrier

In 625/50 PAL, the color subcarrier frequency is based on an odd multiple of one-quarter the line rate, using the factor $^{1135}/_{4}$. The odd multiple of one-quarter, combined with the line-to-line alternation of the phase of the *V* color difference component, causes the *U* and *V* color components to occupy separate parts of the composite signal spectrum. This makes the PAL signal immune to the hue errors that result when an NTSC signal is subject to differential phase distortion.

$$625 \times \left(\frac{50}{2}\,\mathrm{Hz}\right) \times \left(\frac{1135}{4} + \frac{1}{625}\right)$$
$$= 4.433\,618\,750\ \mathrm{MHz}$$

An offset of +25 Hz is added to the PAL subcarrier frequency in order to minimize the visibility of the *Hannover bar* effect. The 25 Hz offset means that the phase relationship of subcarrier to horizontal advances exactly +0.576° each line. Consequently, subcarrier-locked sampling in PAL is not orthogonal: Vertically aligned samples do not have the same subcarrier phase.

$$\frac{1135}{4} + \frac{1}{625} = \frac{709379}{2500} = \frac{11 \times 64489}{2^2 \times 5^3}$$

The introduction of the +25 Hz offset destroyed the simple integer ratio between subcarrier and line rate: The ratio is quite complex, as shown in the margin. The prime factor 64489, transparent to analog systems, is fairly impenetrable to digital techniques.

$4f_{SC}$ sampling

The earliest digital television equipment sampled composite NTSC or PAL video signals. It was convenient for composite digital NTSC equipment to operate at a sampling frequency of exactly four times the color subcarrier frequency, or about 14.318 MHz, denoted $4f_{SC}$. This rate is sometimes loosely referred to as *D–2*, although that term properly refers to the corresponding videotape format, not the sampling structure.

Sampling NTSC at $4f_{SC}$ gives 910 samples per total line (S/TL). A count of 768 samples (3×2^8) encompasses the active samples of a line, including the blanking transitions. A count of 512 (2^9) lines is just slightly more than the number of nonblanked lines in 525/59.94 scanning. The numbers 768 and 512 are convenient for memory systems: 512 is a power of 2, and 768 is 3 times a power of 2. In the early days of digital television, this combination – 768 and 512 – led to very simple memory and addressing circuits for framestores. The importance of this special combination of 768 and 512 is now diminished: Framestore systems often have more than a single frame of memory, memory devices have much higher capacities, and total memory capacity is now a more important constraint than active sample and line counts. In any case, the binary numbers 768 and 512 were never any help in the design of 625/50 framestores.

In a $4f_{SC}$ system, sampling on the $B'-Y'$ and $R'-Y'$ axes is simple because the digital version of the unmodulated color subcarrier contains only the sample values 0, +1, 0, and –1. This choice of sampling axes makes color demodulation a simple process of changing signs. There are severe penalties to this approach, however: The approach is susceptible to *cycle-hopping* when the subcarrier-to-horizontal relationship is unlocked or not carefully controlled, color-under signals cannot be processed, and hue adjustment cannot be implemented without shifting the horizontal picture position.

Any significant processing of a picture, such as repositioning, resizing, rotating, and so on, requires that the signal be represented in components. For this reason, component video equipment is preferred in production and postproduction. But $4f_{SC}$ equipment is cheaper than component equipment, and continues to dominate broadcast operations.

Common sampling rate

The designers of the NTSC and PAL systems chose video parameters based on simple integer ratios. When component digital sampling became feasible it came as a surprise that the ratio of line duration of 525/59.94 and 625/50 systems turned out to be the ratio of 144 to 143, derived as shown in Table 11.1 below.

Table 11.1 **Derivation of 13.5 MHz common sampling rate.**

$f_{H,525/59.94}$	:	$f_{H,625/50}$
$525 \times \frac{60}{2} \times \frac{1000}{1001}$	:	$625 \times \frac{50}{2}$
$7 \times 5^2 \times 3 \times \frac{5 \cdot 3 \cdot 2^2}{2} \times \frac{5^3 \cdot 2^3}{13 \cdot 11 \cdot 7}$	:	$5^4 \times \frac{5^2 \times 2}{2}$
$3 \times 3 \times 2^4$	:	13×11
144	:	143

The lowest common sampling frequency corresponding to these factors is 2.25 MHz, half of the now-familiar NTSC sound subcarrier frequency of 4.5 MHz. Any multiple of 2.25 MHz could have been used as the basis for line-locked sampling of both 525/59.94 and 625/50. The most practical sampling frequency is 6 times 2.25 MHz, or 13.5 MHz; this multiplier is a compromise between a rate high enough to ease the design of analog antialiasing filters, and low enough to minimize data rate and memory requirements.

At 13.5 MHz, 525/59.94 video has 858 samples per total line (S/TL), and 625/50 video has 864 S/TL. Fortunately the blanking tolerances between NTSC and PAL accommodate a choice of 720 samples per active line (S/AL) in both systems. Standardization of the number of active samples results in a high degree of commonality in the design of video processing equipment, since only the difference in active line counts needs to be accommodated to serve both 525 and 625 markets. Also the technically difficult problem of standards conversion is eased somewhat with a common sampling frequency, since horizontal interpolation becomes unnecessary. However, blanking must be treated differently in the two systems to meet studio interchange standards.

Genlock

Different pieces of video equipment are locked together using the *genlock* process. This requires that each piece of processing or recording equipment contain a voltage-controlled crystal oscillator (VCXO). The frequency swing of a VCXO is typically about 100 parts per million, or perhaps 1.5 kHz at 14.318 MHz. This a hundred times better than the tolerance of typical PC video, and ten times better than the tolerance of typical workstation video. Studio equipment cannot usually genlock to a PC or workstation.

525/59.94 scanning and sync 12

This chapter details the scanning, timing, and sync structure of 525/59.94/2:1 video. The scanning and timing information in this chapter applies to all variants of 525/59.94 video, both analog and digital. The sync information relates to component analog, composite analog, and composite digital systems.

Frame rate

$$\frac{30}{1.001} \approx 29.97$$

525/59.94 video represents stationary or moving two-dimensional images sampled temporally at a constant rate of $^{30}/_{1.001}$ frames per second. For studio video, the tolerance on frame rate is normally ±10 ppm. In practice the tolerance applies to a master clock at a high frequency, but for purposes of computation and standards writing it is convenient to reference the tolerance to the frame rate.

Interlace

$$\frac{9}{0.572} \approx 15.734$$

A frame comprises a total of 525 horizontal raster lines of equal duration, uniformly scanned top to bottom and left to right, numbered consecutively starting at 1. Scanning has 2:1 interlace to form an *odd* field and an *even* field; scan lines in the even field are displaced vertically by half the vertical sampling pitch, and delayed temporally by half the frame time, from scanning lines in the odd field.

EQ: Equalization pulse

BR: Broad pulse

CC: Closed Caption

[1]: Line number relative to start of even field

Table 12.1 **525/59.94 line assignment.**

Line number, Odd field	*Line number, Even field*	*Contents, Left half*	*Contents, Right half*
1		EQ	EQ
	264 [1]	EQ	EQ
2		EQ	EQ
	265 [2]	EQ	EQ
3		EQ	EQ
	266 [3]	EQ	BR
4		BR	BR
	267 [4]	BR	BR
5		BR	BR
	268 [5]	BR	BR
6		BR	BR
	269 [6]	BR	EQ
7		EQ	EQ
	270 [7]	EQ	EQ
8		EQ	EQ
	271 [8]	EQ	EQ
9		EQ	EQ
	272 [9]	EQ	none
10–19		Vertical interval video (10 lines)	
	273–282 [10–19]	Vertical interval video (10 lines)	
20		Vertical interval video	
	283 [20]	Vertical interval video	
21		CC or picture	
	284 [21]	CC or picture	CC or Even picture
22–262		Odd picture (241 lines)	
	285–525 [22–262]	Even picture (241 lines)	
263		Odd picture	EQ

The vertical center of the picture is located midway between lines 404 and 142.

Table 12.1 opposite shows the vertical structure of a frame in 525/59.94 video, and indicates the assignment of line numbers and their content.

The Odd field comprises 263 lines:

- Starting on line 1, nine lines of odd vertical sync
- Starting on line 10, twelve lines of vertical interval video
- Starting on line 22, two hundred forty one picture lines
- On line 263, in the left half, picture

The Even field comprises 262 lines:

- Starting on line 264, nine lines of even vertical sync
- Starting on line 273, eleven lines of vertical interval video
- On line 284, in the right half, picture
- Starting on line 285, two hundred forty one picture lines

For details concerning VITC line assignment, see SMPTE RP 164-1992, *Location of Vertical Interval Timecode.*

Vertical interval video lines do not convey picture information, although they may convey other signals either related or unrelated to the picture. If *vertical interval timecode* (VITC) is used, it should be located on line 14 (277), but may be placed on lines 16 and 18 (279 and 281). See *Vertical interval timecode, VITC*, on page 268.

For details concerning closed captions, see Electronic Industries Association ANSI/EIA 608-1994, *Recommended Practice for Line 21 Data Service.*

Upon transmission, picture on lines 21 and 284 may be replaced by line 21 data and/or closed captions.

Horizontal events are referenced to an instant in time denoted 0_H, pronounced *zero-H*. In the analog domain, 0_H is defined by the 50-percent point of the leading (negative-going) edge of each line sync pulse.

In a component digital interface, the correspondence between sync and the digital information is determined by a *timing reference sequence* (TRS) conveyed across the interface.

Vertical events are referenced to an instant in time denoted 0_V, pronounced *zero-vee*. Historically each field was considered to comprise $262\frac{1}{2}$ lines: 0_V for the odd (first) field was defined by the 50 percent point of the first equalization pulse that started the field, and 0_V for the even field was defined by the 50-percent point of the equalization pulse occurring halfway between 0_H instants. In the digital domain it is convenient to consider each field to comprise a whole number of lines. The odd field is usually taken to comprise 263 lines, and the even field to comprise 262 lines. In the digital domain, the important 0_V is the one at the start of the frame.

Although analog terminology considered a field to comprise $262\frac{1}{2}$ lines, the lines were numbered 1 through 263 in the odd field, and 1 through 262 in the even field. In modern practice, line numbering continues through the frame. (In 625/50 systems, line numbering continues through the frame, but starts differently with respect to vertical sync.)

Line sync

In an analog interface, every line commences at 0_H with the negative-going edge of a sync pulse. With the exception of vertical sync lines, which I will describe in a moment, each line commences with a *normal* sync pulse, to be described. Each line that commences with normal sync may contain video information. In composite video, each line that commences with normal sync must contain *burst*, to be described on page 222. Component video has no burst.

Today, even monochrome (black-and-white) video signals have burst. A decoder should accept, as monochrome, any signal without burst.

Every line that commences with a sync pulse other than normal sync maintains blanking level, except for the intervals occupied by sync pulses.

Field/frame sync

To define vertical sync, the frame is divided into intervals of halfline duration. Each halfline either contains no sync information, or commences with the assertion of a sync pulse having one of three durations, each having a tolerance of ±0.100 µs:

EIA and FCC standards in the United States rounded the equalization pulse duration to two digits, to 2.3 µs, slightly less than the theoretical value of 2.35 µs. Equipment is usually designed to the letter of the regulation, not its intent.

- A *normal* sync pulse having a duration of 4.7 µs,
- An *equalization* pulse having half the duration of a normal sync pulse, or
- A *broad* pulse, having a duration of half the line time less the duration of a normal sync pulse.

Each set of 525 halflines in the field commences with a vertical sync sequence as follows:

This sequence causes line 263 to commence with a normal sync pulse and have an equalization pulse halfway through the line. Line 272 commences with an equalization pulse and remains at blanking with no sync pulse halfway through the line.

- Six preequalization pulses,
- Six broad pulses, then
- Six postequalization pulses.

Figure 12.1 overleaf shows details of the sync structure; this waveform diagram is the analog of the table *525/59.94 line assignment*, on page 208.

When sync is represented in analog or digitized form, a raised-cosine transition having a rise time (from 10 percent to 90 percent) of 140 ns ±20 ns is imposed, where the midpoint of the transition is coincident with the idealized sync.

SMPTE RP 168-1993, *Definition of Vertical Interval Switching Point for Synchronous Video Switching.*

It is standard to switch a video signal halfway through the first normal line of the field, line 10 in field 1 and line 273 of field 2 of a 525/59.94 signal. This point is chosen to avoid conflict with sync.

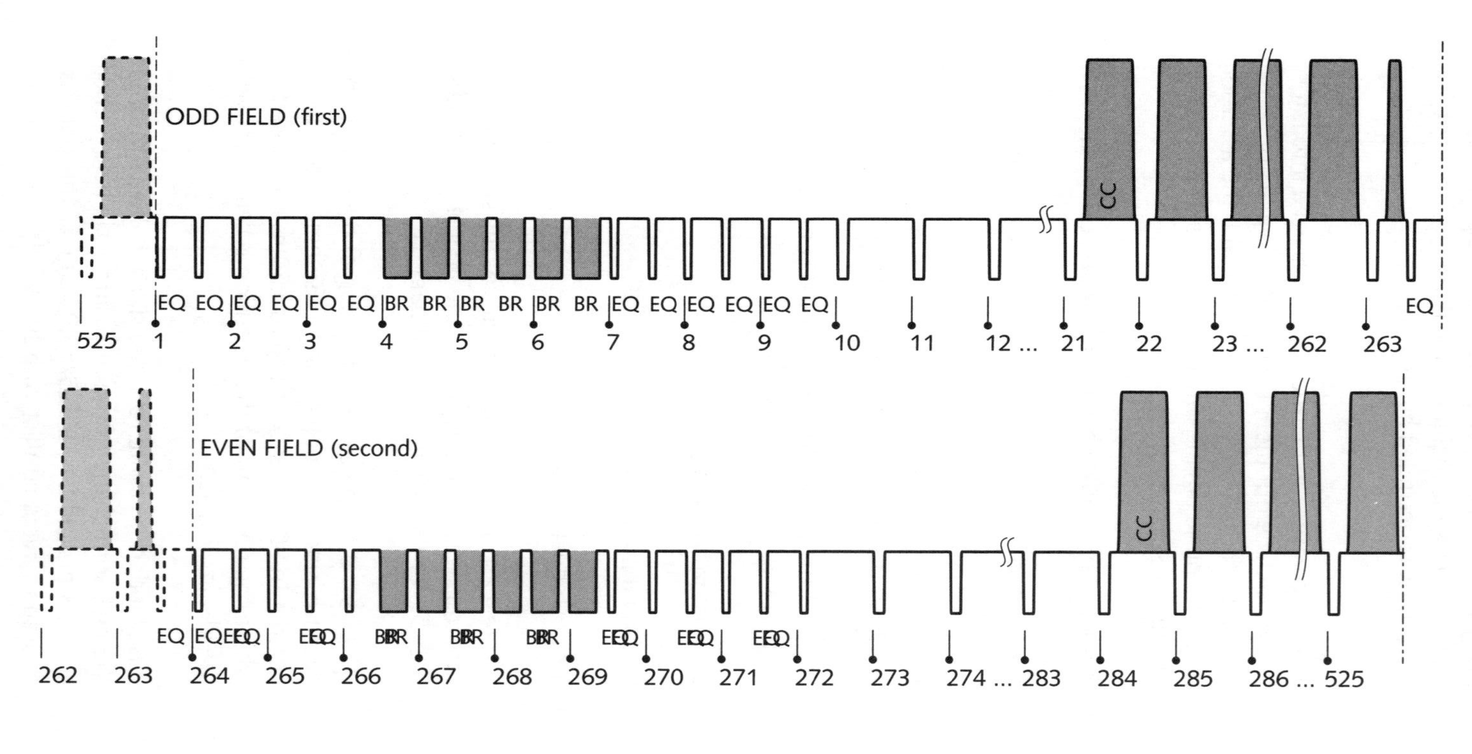

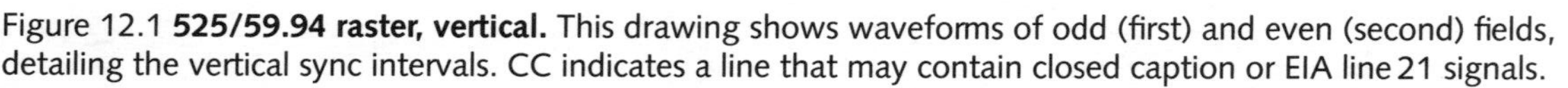
Figure 12.1 **525/59.94 raster, vertical.** This drawing shows waveforms of odd (first) and even (second) fields, detailing the vertical sync intervals. CC indicates a line that may contain closed caption or EIA line 21 signals.

Sync distribution

Although a component video signal *per se* has no burst, it is common to include burst on the reference video signals of component video systems. In other words, both analog and digital component video systems use composite reference signals.

To distribute timing information among component analog video equipment, it is conventional to use a sync pulse signal having an amplitude of either 286 mV or 300 mV. The signal can be considered to be a legal video signal with zero setup whose picture information is entirely black.

In a composite system, a *colorblack* or *blackburst* reference signal is used that comprises sync, blanking, burst, and setup. This signal represents an entirely black picture, with zero luma and zero chroma everywhere, in accordance with *525/59.94 NTSC composite video*, on page 221.

Prior to the widespread adoption of blackburst as a reference video signal, synchronization was accomplished in studio and network facilities by distributing multiple pulse signals, usually having amplitudes of 4 V_{PP} or 2 V_{PP}.

Picture center, aspect ratio, and blanking

SMPTE RP 187, *Center, Aspect Ratio and Blanking of Video Images* (proposal).

The proposed SMPTE RP 187 calls for the center of the picture in 525/59.94 video to be located at the fraction $^{321}/_{572}$ between 0_H instants.

Aspect ratio is defined to be 4:3 with respect to a *clean aperture* pixel array, 708 samples wide at sampling rate of 13.5 MHz, and 480 lines high. Widescreen versions of 525/59.94 have been defined with an aspect ratio of 16:9 instead of 4:3, but otherwise conform to the parameters of standard 525/59.94.

In *Transition samples*, on page 79, I mentioned that it is necessary to avoid, at the start of a line, an instantaneous transition from blanking to picture information. SMPTE standards call for picture information to have a risetime of 140 ns ±20 ns. For 525/59.94 or 625/50 video, a blanking transition is best implemented as a three-sample sequence where the video signal is clamped in turn to 10 percent, 50 percent, and 90 percent of its full excursion. These limits should be applied as clamp (minimum/maximum) values, rather than multipliers, to avoid disturbing the transition samples of a signal that has already had proper blanking applied. A blanking transition should not intrude on the clean aperture; timing details are presented in the following two chapters.

Halfline blanking

Halfline blanking has been abolished from HDTV.

Most component video equipment treats the top and bottom lines of both fields as integral lines; blanking of halflines is assumed to be imposed at the time of conversion to analog. In composite equipment and analog equipment, halfline blanking must be imposed.

$$30.593 \approx \frac{63.556}{2} - \frac{732-716}{13.5}$$

In the composite and analog domains, video information at the bottom of the picture, on the left half of line 263, should terminate 30.593 µs after 0_H. This timing is comparable to blanking at end of a full line, but preceding the midpoint between 0_H instants instead of preceding the 0_H instant itself.

$$41.259 \approx \frac{63.556}{2} - \frac{858-732-2}{13.5}$$

In the composite and analog domains, a right halfline at the top of the picture – such as picture on line 284 – should commence 41.259 µs after 0_H. This timing is comparable to blanking at start of a full line, but following the midpoint between 0_H instants instead of following the 0_H instant itself.

525/59.94 component video 13

This chapter details the coding of component video in 525/59.94 systems. I assume you are familiar with *525/59.94 scanning and sync* on page 207.

RGB primary components

Picture information originates with linear-light primary red, green, and blue (*RGB*) tristimulus components, represented in abstract terms in the range 0 (reference black) to +1 (reference white). In modern standards for 525/59.94, the colorimetric properties of the primary components conform to *SMPTE RP 145 primaries*, described on page 136.

Nonlinear transfer function

From *RGB* tristimulus values, three nonlinear primary components are computed according to the optoelectronic transfer function of *Rec. 709 transfer function*, described on page 102. R denotes a tristimulus component and R'_{709} denotes a nonlinear primary component:

Eq 13.1

$$R'_{709} = \begin{cases} 4.5\,R, & R \le 0.018 \\ 1.099\,R^{0.45} - 0.099, & 0.018 < R \end{cases}$$

This process is loosely called *gamma correction*.

The R', G', and B' components should maintain time-coincidence with each other, and with sync, within ±25 ns.

Luma, Y'

Luma in 525/59.94 systems is computed as a weighted sum of nonlinear R', G', and B' primary components, according to the luma coefficients of Rec. 601:

Eq 13.2

$$Y'_{601} = 0.299\,R' + 0.587\,G' + 0.114\,B'$$

The luma component Y', being a weighted sum of nonlinear $R'G'B'$ components, has no simple relationship with the CIE luminance tristimulus component, denoted Y, used in color science. Video encoding specifications typically place no upper bound on luma bandwidth (though transmission standards may).

Component digital 4:2:2 interface

The C_B and C_R color difference signals of component digital video are formed by scaling B'-Y' and R'-Y' components, as described in *C_BC_R components* on page 174. $Y'C_BC_R$ signals are conveyed according to *Parallel digital interface* on page 248, optionally serialized according to *Serial digital interface* on page 249.

No SMPTE standard addresses square pixel sampling of 525/59.94 video. I recommend using a sample rate of $780f_H$, that is, $12\,^3/_{11}$ MHz. Use 648 samples – or, failing that, 644 or 640 – positioned according to the centerline of Figure 13.2.

In component digital video, the sample structure aligns with 0_H. With 13.5 MHz sampling of 525/59.94, sample 732 corresponds to the horizontal datum (0_H) instant: If digitized, that sample would take the 50 percent value of analog sync.

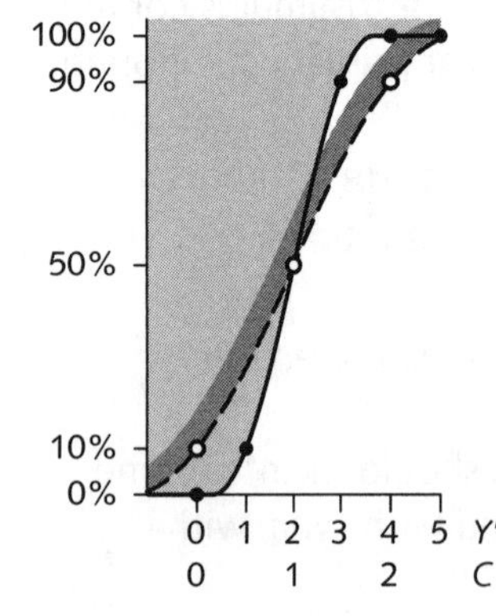

Figure 13.1 **Blanking samples.**

The 4:2:2 interface provides 720 *active* samples, numbered 0 through 719. Sample intervals outside the active area are implicitly blanked. The first few samples of the active region, and the last few, accommodate blanking transition samples. As mentioned in *Picture center, aspect ratio, and blanking*, on page 213, it is recommended to have three transition samples, at 10 percent, 50 percent and 90 percent of the full signal amplitude. Proposed SMPTE RP 187 suggests that samples 2 and 716 should correspond to the 50 percent points of picture width. Figure 13.1 in the margin sketches the transition samples: The dots and light shading represents luma limits; the open circles and heavy shading represents 4:2:2 chroma limits.

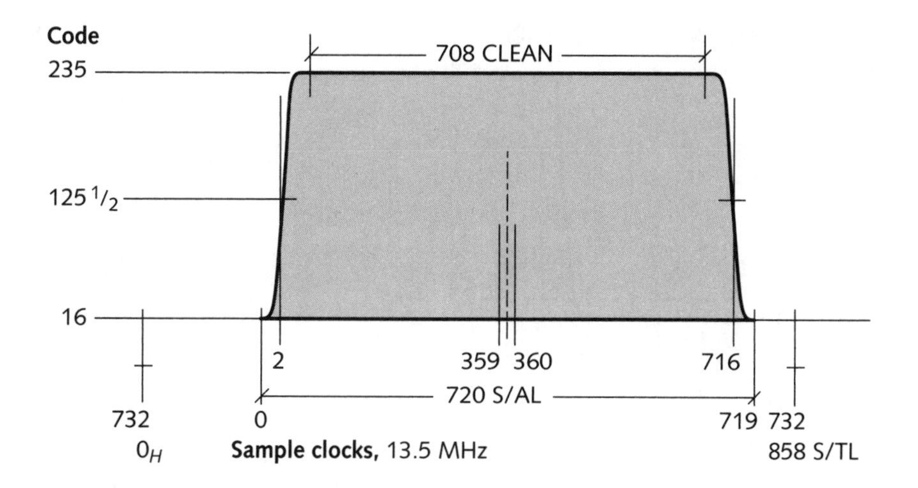

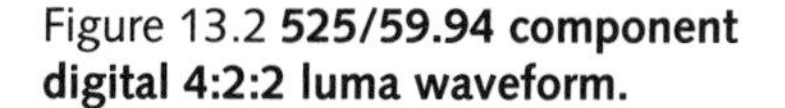

Figure 13.2 **525/59.94 component digital 4:2:2 luma waveform.**

Figure 13.2 above shows a waveform drawing of luma in a 525/59.94 component digital 4:2:2 system.

A derivative of SMPTE 125M uses two channels to convey *R'G'B'* 4:4:4 (or *R'G'B'A* 4:4:4:4) components.

Component analog R'G'B' interface

A component 525/59.94 *R'G'B'* interface is based on nonlinear *R'*, *G'*, and *B'* signals, conveyed according to *Electrical and mechanical interfaces*, on page 247.

In studio systems, analog component *R'G'B'* signals have zero setup, so zero in Equation 13.1 corresponds to 0 V_{DC}. Using SMPTE standards, unity corresponds to 700 mV. Sync is added to the green component according to Equation 13.3, where *sync* and *active* are taken to be unity when asserted and zero otherwise:

Eq 13.3

$$G'_{sync} = \frac{7}{10}\left(active \cdot G'\right) + \frac{3}{10}\left(-sync\right)$$

Some systems, such as 525/59.94 studio video in Japan, use a picture-to-sync ratio of 10:4 and zero

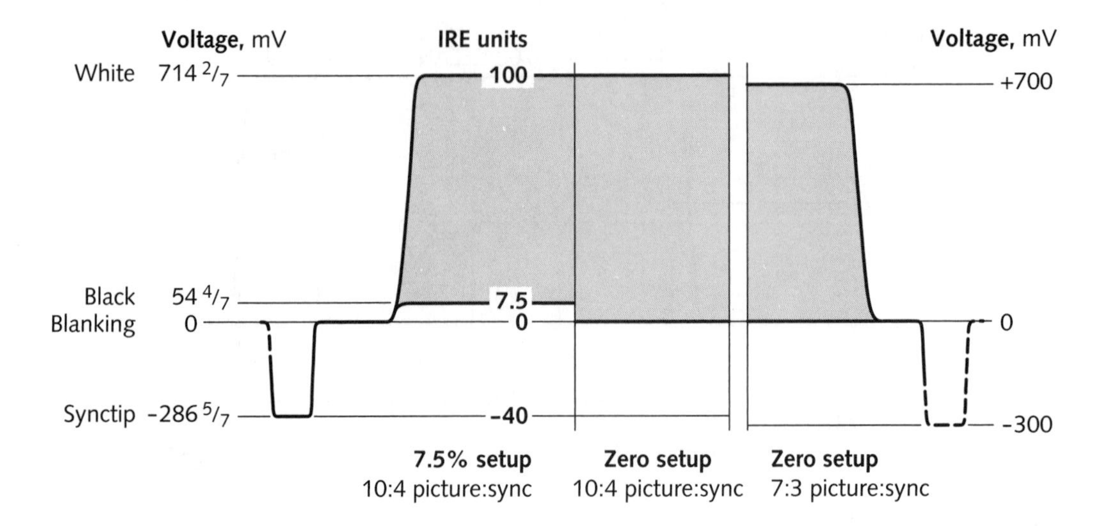

Figure 13.3 **Comparison of 7.5-percent and zero setup.** The left-hand region shows the video levels of composite 525/59.94 video, with 7.5 percent setup and 10:4 picture-to-sync ratio. This coding is used in some studio equipment, and in many computer monitor interfaces. The central region shows zero setup used with 10:4 picture-to-sync; this coding is used in Japan. SMPTE component video and 625/50 systems use zero setup, 700 mV picture, and 300 mV sync, as shown at the right.

setup. In this case, unity in Equation 13.1 corresponds to $5/7$ V, about 714 mV:

Eq 13.4

$$G'_{sync} = \frac{5}{7}\left(active \cdot G'\right) + \frac{2}{7}\left(-sync\right)$$

Finally, some systems – such as computer framebuffers using the levels of the archaic EIA RS-343-A standard – code component video similarly to composite video, with 10:4 picture-to-sync ratio and 7.5-percent setup:

Eq 13.5

$$G'_{sync} = \frac{3}{56} active + \frac{37}{56}\left(active \cdot G'\right) + \frac{2}{7}\left(-sync\right)$$

Figure 13.3 above shows these variations. For further detail, consult *Setup* on page 224.

Use of sync-on-green in computer monitor interfaces was once common, but separate sync is now preferred.

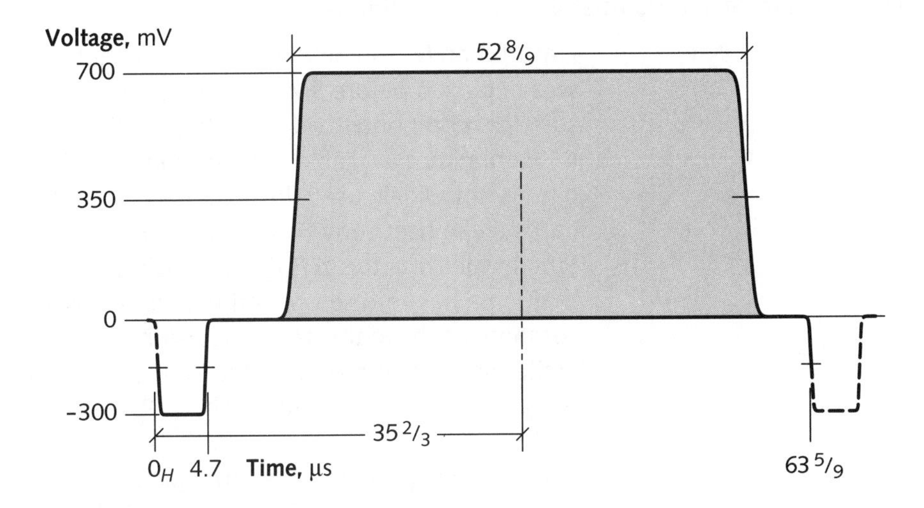

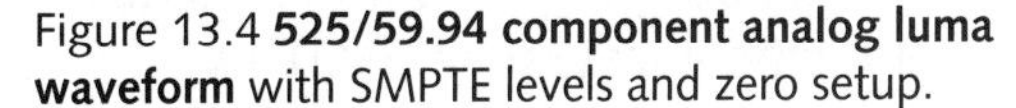

Figure 13.4 **525/59.94 component analog luma waveform** with SMPTE levels and zero setup.

Component analog $Y'P_BP_R$ interface, SMPTE

The P_B and P_R scale factors are appropriate only for component analog interfaces. Consult *C_BC_R components*, on page 174, for details concerning scale factors for component digital systems; or *UV components*, on page 180, for details concerning scale factors for composite analog or digital NTSC or PAL.

The P_B and P_R color difference signals of component analog video are formed by scaling B'-Y' and R'-Y' components, as described in *P_BP_R components* on page 173. Although it is possible in theory to have wideband P_B and P_R components, in practice they are lowpass filtered to about half the bandwidth of luma.

Y', P_B, and P_R signals are conveyed electrically according to *Electrical and mechanical interfaces* on page 247.

Component $Y'P_BP_R$ signals, according to the SMPTE standard, employ zero setup. Zero (reference blanking level) for Y', P_B, and P_R corresponds to a level of 0 V_{DC}, and unity corresponds to 700 mV. Sync is added to the luma component; *sync* is taken to be unity when asserted and zero otherwise:

Eq 13.6

$$Y'_{sync} = \frac{7}{10} Y'_{601} + \frac{3}{10}(-sync)$$

Figure 13.4 above shows a waveform drawing of luma in a 525/59.94 component analog interface according to the proposed SMPTE 253M standard.

Component analog $Y'P_BP_R$ interface, industry standard

Unfortunately, equipment from two manufacturers was deployed before SMPTE reached agreement on a standard component video analog interface. Although it is sometimes available as an option, the SMPTE standard is not widely used. Instead, two "industry" standards are in use: Sony BetaCam and Panasonic M-II. Strictly speaking the $Y'P_BP_R$ nomenclature indicates that luma has zero setup, and that color difference components have the same excursion (from black to white) as luma. But both of the industry standards use setup, and neither adheres to the $Y'P_BP_R$ convention.

BetaCam equipment utilizes 10:4 picture-to-sync ratio (roughly 714 mV luma, 286 mV sync) with 7.5 percent setup on luma. Color differences range ±350 mV.

M-II equipment utilizes 7:3 picture-to-sync ratio (exactly 700 mV luma, 300 mV sync) with 7.5 percent setup on luma. Color differences are scaled by the $^{37}/_{40}$ setup fraction, for an excursion of ±324.5 mV.

525/59.94 NTSC composite video 14

Although this book is mainly concerned with *digital* video, the installed base of hundreds of millions of units of analog video equipment cannot be ignored. Furthermore, composite $4f_{SC}$ digital video – or, loosely, *D-2* – is essentially just digitized analog video. To understand $4f_{SC}$ equipment, you must understand conventional analog NTSC video.

This chapter details the technical parameters of composite 525/59.94 video: subcarrier, composite chroma, composite analog NTSC, and the S-video-525 interface. I assume that you are familiar with the concepts of *Composite NTSC and PAL*, on page 185; with the sync structure detailed in *525/59.94 scanning and sync*, on page 207; and with *R'G'B'* coding described in *525/59.94 component video*, on page 215. Before I explain the signal flow, I must introduce subcarrier and burst.

Subcarrier

Synchronous with the scanning raster is a pair of continuous-wave subcarriers having 227 ½ cycles per total raster line: a sine wave (referred to as *subcarrier*) whose zero-crossing is coincident ±10° with 0_H, and a cosine wave in quadrature (at 90°). Delay of the subcarrier's zero-crossing with respect to the 0_H datum, measured in degrees of subcarrier, is known as *subcarrier to horizontal* (SCH) error.

Derived color subcarrier frequency is $^{315}/_{88}$ MHz ± 10 ppm, or about 3.58 MHz. Subcarrier drift should not exceed $\pm ^{1}/_{10}$ Hz per second. Subcarrier jitter should not exceed ± 0.5 ns over one line time.

Two-frame sequence

$$\frac{455}{2} \times 525 = 119\,437.5$$

Because the total number of subcarrier cycles per line is an odd multiple of one-half, and there are an odd number of lines per frame, subcarrier can fall in one of two relationships with the start of a frame. *Colorframes* denoted *A* and *B* are distinguished by the phase of subcarrier at 0_H at the start of the frame: *Frame A* at 0°, and *Frame B* at 180°. This relationship is also referred to as a four-field sequence of fields 1, 2, 3, and 4 (or I, II, III, and IV), corresponding to A_{odd}, A_{even}, B_{odd}, B_{even}.

Burst

The inversion of burst from subcarrier puts burst at 180° on a vectorscope display.

It is unfortunate that the standards are written to have cumulative tolerances at the end of the burst envelope.

Burst is formed by multiplying the inverted *sin* subcarrier by a *burst gate* that is asserted 19 subcarrier cycles $^{+200}_{-100}$ ns after 0_H on every line that commences with a normal sync pulse. Burst gate has a duration of 9 ± 1 subcarrier cycles. Burst gate has raised-cosine transitions whose 50 percent points are coincident with the time intervals specified above, and whose risetimes are 300^{+200}_{-100} ns.

Figure 14.1 below sketches the relationship of subcarrier, sync, and burst. Subcarrier may be in the phase indicated, or inverted, depending on colorframe.

Figure 14.1 **Subcarrier to horizontal (SCH) relationship.**

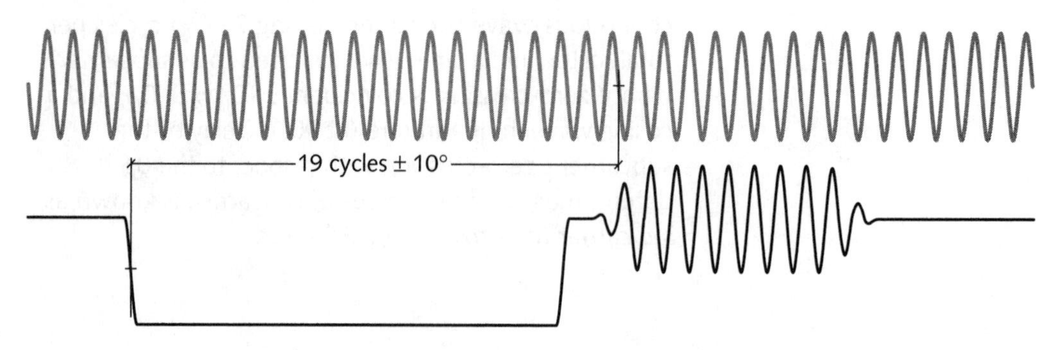

Color differences, U, V

As described in *UV components*, on page 180, color differences for S-video and NTSC are computed by scaling $B'-Y'$ and $R'-Y'$ components to form U and V. This scaling limits the maximum value of the composite signal, to be defined in *Composite NTSC encoding*, on page 225, to the range $-\frac{1}{3}$ to $+\frac{4}{3}$. The scale factors would cause 100 percent colorbars to have an excursion from $-33\frac{1}{3}$ IRE to $+133\frac{1}{3}$ IRE, were it not for the scaling by $\frac{37}{40}$ that occurs when setup is introduced. See *Setup* on page 224.

The $+\frac{4}{3}$ limit applies to composite studio equipment; a limit of +1.2 applies to terrestrial (VHF/UHF) broadcast.

Color difference filtering

The U and V color difference components are subjected to lowpass filters having attenuation:

- Less than 2 dB at frequencies less than 1.3 MHz
- At least 0 dB between 1.3 MHz and 3.6 MHz
- At least 20 dB at frequencies greater than 3.6 MHz

Chroma, C

As explained in *Quadrature modulation*, on page 187, U and V color difference components are combined into a single *chroma* signal:

Eq 14.1

$$C = U(\cos t) + V(\sin t)$$

$\sin t$ and $\cos t$ represent the 3.58 MHz color subcarrier defined in *Subcarrier*, on page 221.

To be compliant with FCC regulations for broadcast, an NTSC modulator is supposed to operate on I and Q components, where the Q component is bandwidth limited more severely than the I component:

Eq 14.2

$$C = Q\cos(t + 33°) + I\sin(t + 33°)$$

The bandwidth limitation of Q, to about 600 kHz, was specified in the original design of NTSC to permit accurate recovery of the wideband *I* component. *Y'IQ* was important in the early days of NTSC, but contemporary NTSC equipment modulates equiband *U* and *V*, and this practice has been enshrined in SMPTE 170M. See *Narrowband I chroma*, on page 190.

The *Y'* and *C* components should maintain time-coincidence with each other, and with sync, within ±25 ns. Errors in relative timing are known as *chroma-luma delay*.

Setup

525/59.94 video in Japan employs zero setup.

Composite video signals that use 525/59.94 scanning usually employ *7.5-percent setup*, whereby luma is scaled and offset so that reference white remains at unity, but reference black is raised by the fraction 7.5 percent of full scale. The luma signal remains at zero – *reference blanking level* – during signal intervals when no picture is being conveyed.

When setup is employed, the same scale factor that is applied to luma must also be applied to chroma, to ensure that the peak chroma excursion remains at $+\frac{4}{3}$ of reference white, or, in other words, to ensure that the peaks of 75 percent colorbars exactly match reference white at 100 percent (100 IRE):

Eq 14.3

$$
\begin{aligned}
Y'_{601,setup} &= \frac{3}{40} + \frac{37}{40} Y'_{601} \\
C_{setup} &= \frac{37}{40} C
\end{aligned}
$$

Setup was invented in the vacuum tube era to prevent retrace lines from becoming visible due to unstable black levels. It now serves no useful purpose but is retained for compatibility reasons.

In addition to specifying 7.5-percent setup, RS-343-A specifies monochrome operation, 60.00 Hz field rate, 2:1 interlace, 7 µs horizontal blanking, and other parameters that have no place in modern video systems.

Setup is specified in the archaic EIA RS-343-A standard; consequently, a computer *R'G'B'* interface that advertises *EIA RS-343-A levels* employs setup.

Setup causes problems in maintaining accurate black level reproduction; consequently, setup has been abolished from modern video systems. *Zero setup* is a feature of SMPTE-standard component video, studio video in Japan, all variants of 625/50 video, all variants of HDTV, and high-end workstations.

S-video-525, Y'/C3.58

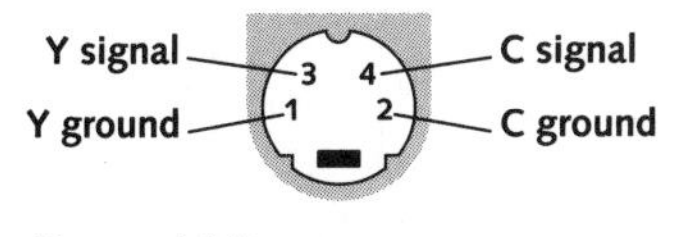

Figure 14.2
S-video connector looking into cable.

An S-video interface conveys luma and chroma separately in order to avoid the cross-chroma and cross-luma artifacts that result from their being summed to form a single NTSC composite signal. The Y' and C signals at an S-video-525 interface have structure and levels identical to the constituent signals of analog NTSC: If the two signals are summed, a legal NTSC signal results.

There are two versions of S-video. *S-video-525* has 525/59.94 scanning, 3.58 MHz color subcarrier, and 7.5-percent setup:

Eq 14.4

$$\begin{aligned} Y'_{S\text{-}video\text{-}525} &= \frac{5}{7} Y'_{601,setup} &+ \frac{2}{7}(-sync) \\ C_{S\text{-}video\text{-}525} &= \frac{5}{7} C_{setup} &+ \frac{2}{7}\left(\frac{burst}{2}\right) \end{aligned}$$

S-video-625, Y'/C4.43, is used with 625/50 scanning; it is detailed on page 243. Thankfully, there is no S-video-PAL-M, S-video-PAL-N, S-video-525-4.43, or S-video-SECAM!

Composite NTSC encoding

A composite NTSC signal is formed by summing luma (Y') and modulated chroma (C) signals, along with sync and burst. In the following expressions, *sync* is taken to be unity when asserted and zero otherwise; *burst* is in the range ±1:

Eq 14.5

$$NTSC_{setup} = \frac{5}{7}\left(Y'_{601,setup} + C_{setup}\right) \quad + \frac{2}{7}\left(\frac{burst}{2} - sync\right)$$

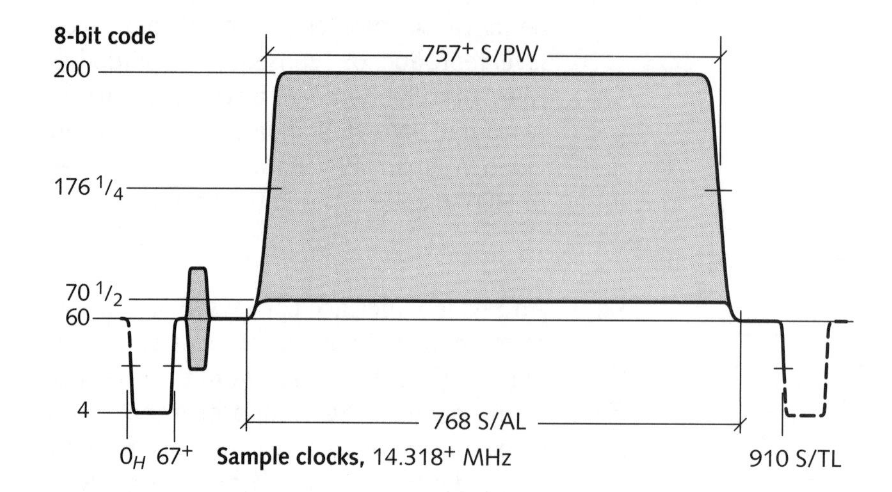

Figure 14.3 **525/59.94 composite digital waveform** with setup. The excursion indicated reflects the range of the luma component; the chroma component will contribute to excursions outside this range.

The excursion of a 525/59.94 video signal with zero chroma – from synctip to reference white – ranges from -40 IRE to +100 IRE. The excursion of maximum chroma is about 131 IRE.

Composite digital interface, $4f_{SC}$

Digital composite NTSC – also known as the 525/59.94 version of $4f_{SC}$ – is formed by scaling and offsetting Equation 14.5 so that in an 8-bit system blanking (zero in the equation) is at codeword 60 and reference white is at codeword 200. In a 10-bit system, the reference codes are multiplied by four: The reference codes remain the same except for having two low-order zero bits appended. Codewords having the 8 most-significant bits all-one or all-zero are prohibited from video data. Figure 14.3 above shows a waveform drawing of luma in a 525/59.94 composite digital interface, with 7.5-percent setup. Sampling is performed on the *I* and Q axes, not *U* and *V*: The 0_H datum does not correspond exactly to a sample instant; it is, in effect, located at sample 784 $^{33}/_{90}$.

SMPTE 244M-1995, *System M/NTSC Composite Video SIgnals – Bit-Parallel Digital Interface.*

The electrical and mechanical interface for 525/59.94, $4f_{SC}$ is derived from the Rec. 656 parallel and serial interfaces; see page 248.

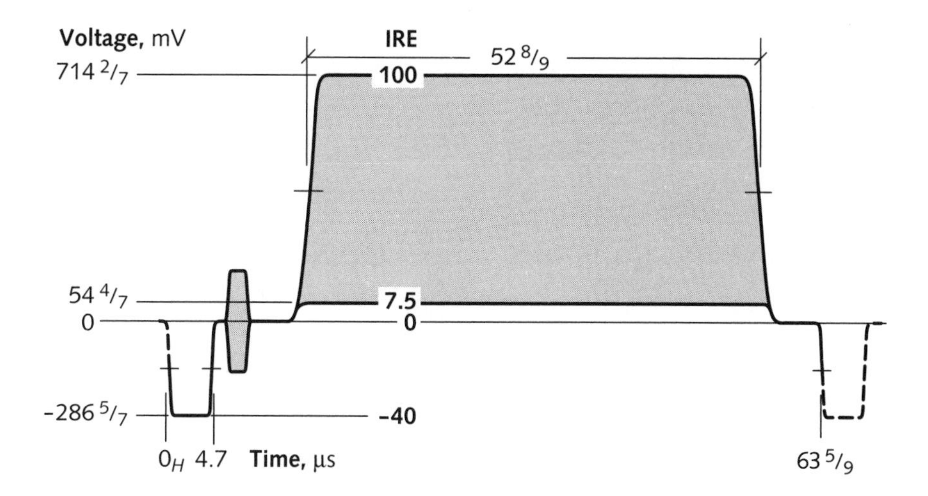

Figure 14.4 **525/59.94 composite analog waveform** showing the luma component, with setup.

Composite analog NTSC interface

SMPTE 170M-1994, *Composite Analog Video SIgnal – NTSC for Studio Applications.*

Figure 14.4 above shows a waveform drawing of luma in a 525/59.94 composite analog interface, with 7.5-percent setup; the levels are detailed in *Analog electrical interface* on page 247. The excursion of a 525/59.94 luma signal – from synctip to reference white – is normally 1 V_{PP}. Including maximum chroma excursion to about 131 IRE, the maximum excursion of the composite NTSC signal is about 1.22 V_{PP}.

625/50 scanning and sync 15

This chapter details the scanning, timing, and sync structure of 625/50/2:1 video. The scanning and timing information here applies to all variants of 625/50 video, both analog and digital. The sync information relates to component analog, composite analog, and composite digital systems. I assume that you are familiar with *525/59.94 scanning and sync* on page 207; this chapter concentrates on the differences.

Frame rate

625/50 video represents stationary or moving two-dimensional images sampled temporally at a constant rate of 25 frames per second. For studio video, the tolerance on frame rate is normally ± 4 ppm. In practice the tolerance applies to a master clock at a high frequency, but for purposes of computation and standards writing it is convenient to reference the tolerance to the frame rate.

Interlace

Derived line rate is 15.625 kHz.

A frame comprises a total of 625 horizontal raster lines of equal duration, uniformly scanned top to bottom and left to right, numbered consecutively starting at 1, with 2:1 interlace to form an *odd* field and an *even* field. Scanning lines in the even field are displaced vertically by half the vertical sampling pitch, and delayed temporally by half the frame time, from scanning lines in the odd field.

EQ: Equalization pulse

BR: Broad pulse

Table 15.1 **625/50 line assignment.**

Line number, Even field	*Line number, Odd field*	*Contents, Left half*	*Contents, Right half*
311		EQ	EQ
	624	EQ	EQ
312		EQ	EQ
	625	EQ	EQ
313		EQ	BR
	1	BR	BR
314		BR	BR
	2	BR	BR
315		BR	BR
	3	BR	EQ
316		EQ	EQ
	4	EQ	EQ
317		EQ	EQ
	5	EQ	EQ
318		EQ	none
	6	Vertical interval video	
319		Vertical interval video	
	21	ITS (optional)	
334		ITS (optional)	
	22	Quiet	
335		Quiet	
	23	none	Even picture
336–622 (287 lines)		Odd picture	
	24–310 (287 lines)	Even picture	
623		Odd picture	EQ

Table 15.1 opposite shows the vertical structure of a frame in 625/50 video, and indicates the assignment of line numbers and their content.

The Odd field comprises 312 lines:

- Starting on line 1, six lines of odd vertical sync
- Starting on line 6, seventeen lines of vertical interval video
- On line 23, in the right half, picture
- Starting on line 24, two hundred eighty seven picture lines
- On lines 311 and 312, two lines commencing even vertical sync

The Even field comprises 313 lines:

- Starting on line 313, seven lines of even vertical sync
- Starting on line 319, seventeen lines of vertical interval video
- Starting on line 336, two hundred eighty seven picture lines
- On line 623, in the left half, picture
- On lines 624 and 625, two lines commencing odd vertical sync

There are several differences from 525/59.94 scanning. There are five pre-equalization, broad, and post-equalization pulses (instead of six of each). The frame is defined to start with the field containing the top line of the picture, actually a right-hand halfline. In contrast, in 525/59.94 scanning, the first picture line of a frame is a full line, and the right-hand halfline at the top of the picture is in the even field.

In 625/50 systems, lines are numbered starting with the first broad sync pulse: preequalization pulses are counted at the end of one field instead of the beginning of the next. This could be considered to be solely a nomenclature issue, but because line numbers are sometimes recorded in binary form as part of a digital video signal, the issue is substantive. In 625/50 systems, lines are always numbered throughout the frame, where 525/59.94 nomenclature sometimes numbers the lines of each field starting at 1.

Horizontal events are referenced to an instant in time denoted 0_H. In the analog domain, 0_H is defined by the 50-percent point of the leading (negative-going) edge of each line sync pulse. In a component digital interface, the correspondence between sync and the digital information is determined by a *timing reference signal* (TRS) conveyed across the interface.

Line sync

In an analog interface, every line commences at 0_H with the negative-going edge of a sync pulse. With the exception of the vertical sync lines of each field, each line commences with the assertion of a *normal* sync pulse, to be described. Each line that commences with a sync pulse *other* than normal sync maintains blanking level, here denoted zero, except for the interval(s) occupied by sync pulses.

Field/frame sync

Line 1 is defined by the first broad pulse coincident with 0_H.

To define vertical sync, the frame is divided into intervals of halfline duration. Each halfline either contains no sync information, or commences with the assertion of a sync pulse having one of three durations, each having a tolerance of ±0.100 μs:

- A *normal* sync pulse having a duration of 4.7 μs,
- An *equalization* pulse having half the duration of a normal sync pulse, or
- A *broad* pulse having a duration of half the line time less the duration of a normal sync pulse.

Each set of 625 halflines in the frame is associated with a vertical sync sequence, as follows:

This sequence causes line 623 to commence with a normal sync pulse and have an equalization pulse halfway through the line. Line 318 commences with an equalization pulse and remains at blanking with no sync pulse halfway through the line.

- Five preequalization pulses,
- Five broad pulses, then
- Five postequalization pulses.

Figure 15.1 overleaf shows the vertical sync structure of 625/50 analog video. This waveform diagram is the analog of *625/50 line assignment* on page 230.

When sync is represented in analog or digitized form, a raised-cosine transition having a risetime (from 10 percent to 90 percent) of 200 ns ±20 ns is imposed, where the midpoint of the transition is coincident with the idealized sync.

SMPTE RP 168-1993, *Definition of Vertical Interval Switching Point for Synchronous Video Switching.*

It is standard to switch a video signal halfway through the first normal line of each field, lines 6 and 319 of a 625/50 signal. This point is chosen to avoid conflict with sync.

Sync distribution

To distribute timing information among component analog video equipment, it is conventional to use a sync pulse signal with an amplitude of 300 mV. The signal can be considered to be a legal video signal whose picture information is entirely black.

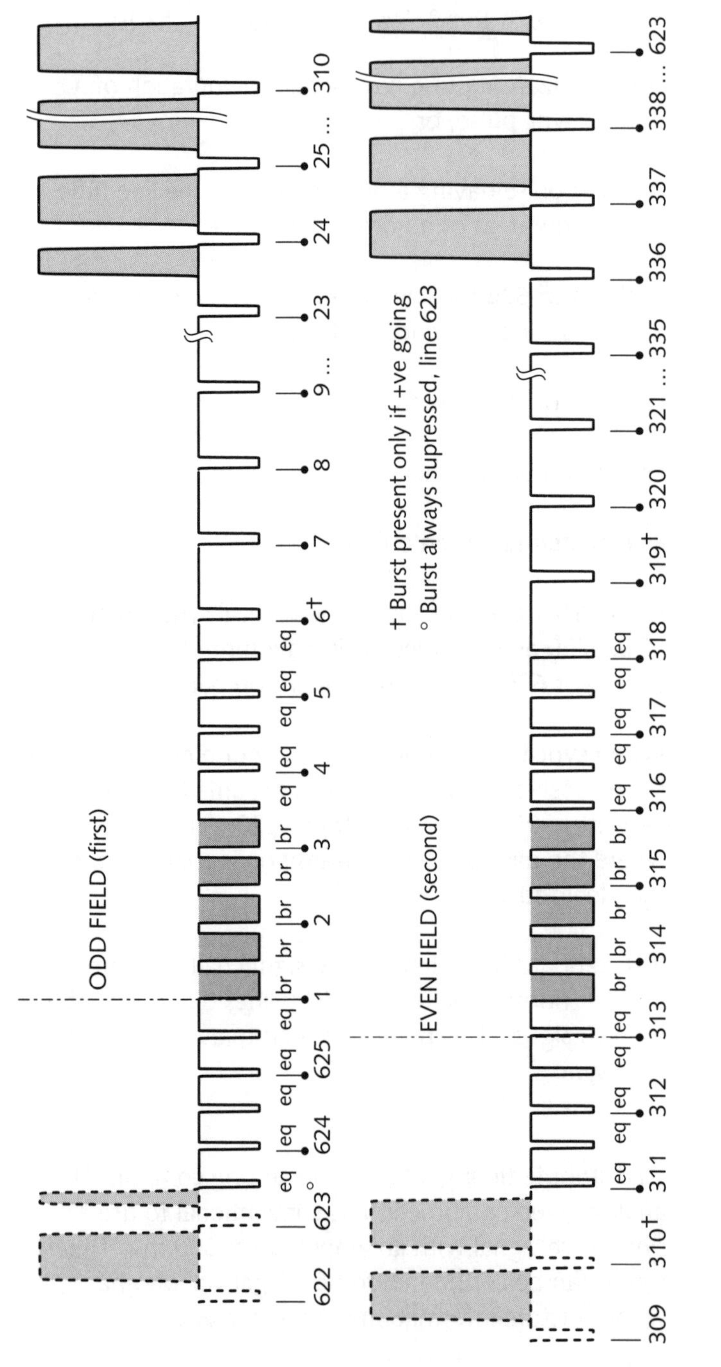

Figure 15.1 **625/50 raster, vertical.** This drawing shows waveforms of odd (first) and even (second) fields, detailing vertical sync intervals. The odd field comprises 312 lines, and the even field comprises 313 lines.

In a composite system, a *color black* or *black burst* reference signal is used, comprising sync, blanking, and burst. This signal represents an entirely black picture, with zero luma and zero chroma everywhere, in accordance with *625/50 PAL composite video* on page 241.

Aspect ratio

The proposed SMPTE RP 187 calls for the center of the picture in 625/50 video to be located at the fraction $^{983}/_{1728}$ between 0_H instants.

Aspect ratio is defined as 4:3 with respect to a *clean aperture* pixel array, 690 samples wide at sampling rate of 13.5 MHz, and 566 lines high. Blanking transitions should not intrude into the clean aperture.

In the composite and analog domains, video information on the left-hand halfline of line 623 terminates 30.350 ±0.1 µs after 0_H. Video information on the right-hand halfline of line 23 commences 42.500 ±0.1 µs after 0_H.

625/50 component video 16

This chapter details 625/50 component video. I assume you are familiar with *625/50 scanning and sync*, detailed on page 229. I will describe 625/50 by explaining the differences from 525/59.94, so I assume that you are quite familiar with *525/59.94 scanning and sync* on page 207.

RGB primary components

Picture information originates with linear-light primary red, green, and blue (*RGB*) tristimulus components, represented in abstract terms in the range 0 (reference black) to +1 (reference white). In modern standards for 625/50, the colorimetric properties of the primary components conform to *EBU primaries* on page 134.

Nonlinear transfer function

625/50 standards documents indicate a precorrection of $1/2.8$, approximately 0.36, but values closer to 0.45 are usually used. See *Gamma in video* on page 100.

From *RGB* tristimulus values, three nonlinear primary components are computed according to the optoelectronic transfer function of Rec. 709, where R denotes a tristimulus component, and R'_{709} denotes a nonlinear primary component:

Eq 16.1

$$R'_{709} = \begin{cases} 4.5R, & R \le 0.018 \\ 1.099\,R^{0.45} - 0.099, & 0.018 < R \end{cases}$$

This process is loosely called *gamma correction*.

The R', G', and B' components should maintain time-coincidence with each other, and with sync, within ±25 ns.

Luma, Y'

Luma in 625/50 systems is computed as a weighted sum of nonlinear R', G', and B' primary components according to the luma coefficients of Rec. 601:

Eq 16.2

$$Y'_{601} = 0.299\,R' + 0.587\,G' + 0.114\,B'$$

The luma component Y', being a weighted sum of nonlinear $R'G'B'$ components, has no simple relationship with the CIE luminance tristimulus component, denoted Y, used in color science. Video encoding specifications typically place no upper bound on luma bandwidth (though transmission standards may).

Component digital 4:2:2 interface

The choice of 720 active samples for Rec. 601 accommodates the blanking requirements of both 525/59.94 and 625/50 analog video: 720 samples are sufficient to accommodate the necessary transition samples for either system. See page 79.

Unfortunately, the blanking tolerances between 525/59.94 and 625/50 do not permit a *single* choice of blanking transition samples: The narrowest possible picture width in 525/59.94 is several samples too wide to meet 625/50 tolerances.

The C_B and C_R color difference signals of component digital video are formed by scaling B'-Y' and R'-Y' components, as described in *C_BC_R components* on page 174. $Y'C_BC_R$ signals are conveyed according to *Parallel digital interface* on page 248, optionally serialized according to *Serial digital interface* on page 249.

The sample structure aligns with 0_H. In 13.5 MHz sampling of 625/50, sample 736 corresponds to the horizontal datum, 0_H. If digitized, that sample would take the 50 percent value of analog sync. Proposed SMPTE RP 187 suggests that samples 8 and 710 should correspond to the 50 percent points of picture width.

Figure 16.1 opposite shows a waveform drawing of luma in a 625/50 component digital 4:2:2 system.

Component analog R'G'B' interface

The interface for analog R', G', and B' signals is described in *Electrical and mechanical interfaces*, on page 247. Zero (reference blanking level) in the R', G', and B' expressions corresponds to a level of 0 V_{DC}, and unity corresponds to 700 mV.

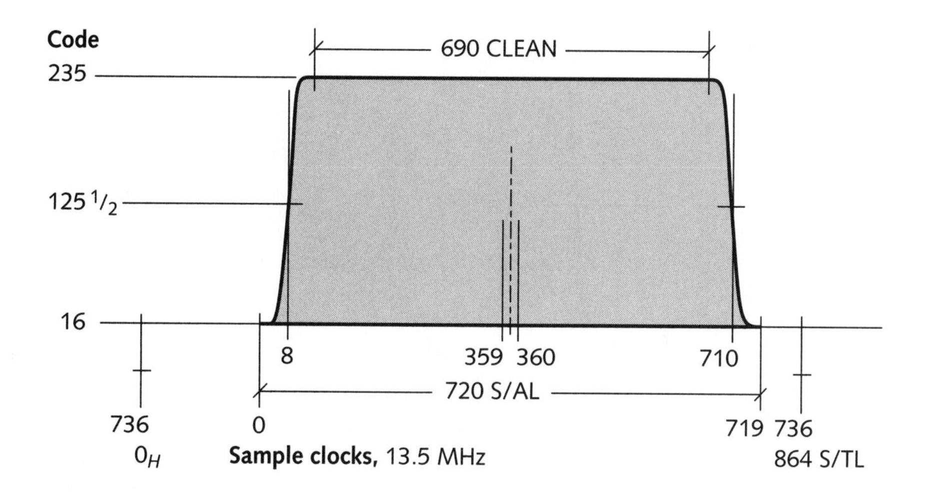

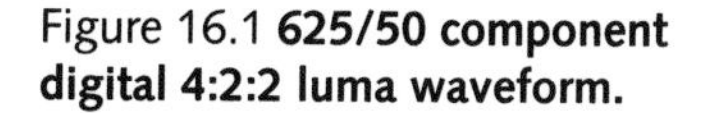

Figure 16.1 **625/50 component digital 4:2:2 luma waveform.**

Sync is added to the green component according to:

Eq 16.3

$$G'_{sync} = \frac{7}{10}\left(active \cdot G'\right) + \frac{3}{10}\left(-sync\right)$$

In component analog video, the excursion of the G' signal from synctip to reference white is 1 V_{PP}. Levels in 625/50 systems are usually specified in millivolts, not the IRE units common in 525/59.94 systems.

Component analog Y'P$_B$P$_R$ interface

The P_B and P_B color difference signals of component analog video are formed by scaling B'-Y' and R'-Y' components. Although it is possible in theory to have wideband P_B and P_R components, in practice they are lowpass filtered to about half the bandwidth of luma.

The interface for analog Y', P_B, and P_R signals is described in *Electrical and mechanical interfaces* on page 247. Component 625/50 $Y'P_BP_R$ signals have zero setup. Sync is added to the luma component according to:

Eq 16.4

$$Y'_{sync} = \frac{7}{10}Y'_{601} + \frac{3}{10}\left(-sync\right)$$

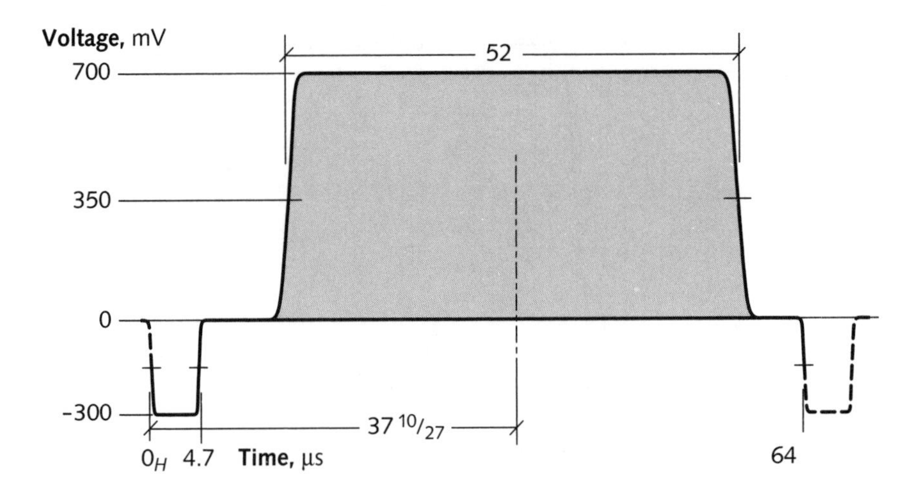

Figure 16.2 **625/50 component analog luma waveform.**

In analog 625/50 component interfaces, this excursion is conveyed as 1 V_{PP}, with reference blanking at 0 V_{DC}. The picture excursion of the Y' signal is 700 mV. Figure 16.2 above shows a waveform drawing of luma in a 625/50 component analog interface.

625/50 PAL composite video 17

This chapter details the formation of 625/50 PAL composite video and S-video-625. I assume that you are familiar with *625/50 scanning and sync* on page 229; and *625/50 component video* on page 237. I describe 625/50 PAL by explaining its differences from 525/59.94 NTSC, so I assume that you are quite familiar with *525/59.94 NTSC composite video* on page 221.

Subcarrier

$$\frac{1135}{4} + \frac{1}{625} = \frac{709379}{2500}$$

$$= 283.7516$$

Synchronous with the 625/50 raster is a pair of continuous-wave subcarriers having exactly 283.7516 cycles per total raster line: a sine-wave (hereafter referred to as *subcarrier*) whose zero-crossing is coincident ±10° with 0_V, and a cosine-wave in quadrature (at 90°). Derived color subcarrier frequency is 4.43361875 MHz ±10 ppm. Subcarrier drift should not exceed ±$^1/_{10}$ Hz per second. Subcarrier jitter should not exceed ±0.5 ns over one line time.

Four-frame sequence

In PAL, the total number of subcarrier cycles per frame is an odd multiple of one-quarter. This causes subcarrier to fall in one of four relationships with the start of a frame. Where necessary, *color frames* are distinguished by the phase of subcarrier at 0_V at the start of the frame. The four-frame sequence is due to the $^{1135}/_4$ fraction, which relates subcarrier to line rate; the fraction $^1/_{625}$ contributes precisely one cycle per frame, so it has no effect on the four-frame sequence.

Burst

Burst – or *colorburst* – is formed by multiplying the inverted *sin* subcarrier by a *burst gate* that is asserted to unity, 5.6 ±0.1 µs after 0_H on every line that commences with a normal sync pulse, and has a duration of 10±1 cycles of subcarrier. Burst gate has raised-cosine transitions whose 50 percent points are coincident with the time intervals specified above, and whose risetimes are 300^{+200}_{-100} ns.

The inversion of burst from subcarrier, combined with the *V*-axis switch, puts PAL burst at +135° and -135° (+225°) on a vectorscope display. This is known as *swinging* burst.

PAL systems have a burst-blanking *meander* scheme: Burst is suppressed from the first and last full lines of a field if it would take -135° phase. Burst is always suppressed from line 623. The suppression of burst is also known as *PAL burst* or *Bruch burst*. The scheme ensures that the first burst immediately preceding and following the vertical interval is always of +135° phase. This provides a PAL decoder with means to recover the *V*-axis switch polarity.

Color difference components, U, V

Color differences for PAL are computed by scaling *B'-Y'* and *R'-Y'* components to form *U* and *V* components, as described on page 180. The scaling limits the maximum value of the composite signal, to be defined in *Composite PAL encoding*, on page 244, to the range -1⁄3 to +4⁄3. The scale factors cause 100 percent colorbars to have an excursion from -33 1⁄3 percent to +133 1⁄3 percent of the picture excursion.

The +4⁄3 limit applies to composite studio equipment; a limit of +1.2 applies to terrestrial (VHF/UHF) broadcast.

Color difference filtering

The U and V color difference components are subjected to lowpass filters having attenuation:

- Less than 3 dB at frequencies less than 1.3 MHz
- At least 0 dB between 1.3 MHz and 4 MHz
- At least 20 dB at frequencies greater than 4 MHz

PAL was standardized with a higher bandwidth than NTSC, so there was no need to severely bandlimit one of the color difference components: PAL uses equiband *U and V* color differences.

Chroma, C

In PAL, the U and V color difference components are combined to form *chroma* signal as follows:

Eq 17.1

$$C = U(\cos t) \pm V(\sin t)$$

sin t and cos t represent the 4.43 MHz color subcarrier defined in *Subcarrier*, on page 241. The process is described in *Quadrature modulation*, on page 187, with the additional feature that the sign of the V component switches *Phase* (±1) on *Alternate Lines*. This is the origin of the acronym *PAL*.

The Y' and C components should be time-coincident with each other, and with sync, within ±25 ns.

S-video-625, Y'/C4.43

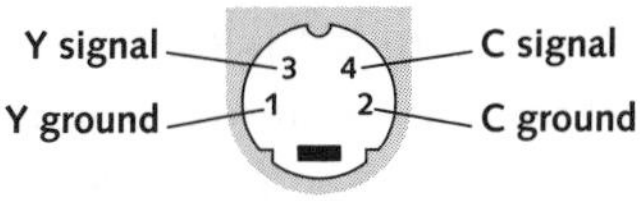

Y signal, Y ground, C signal, C ground (pins 3, 1, 4, 2)

Figure 17.1 **S-video connector** looking into cable.

S-video-625 has 625/50 scanning, 4.43 MHz color subcarrier, and zero setup. The Y' and C signals at the interface have structure and levels identical to the constituent signals of PAL: If the two signals are summed, a legal PAL signal results:

Eq 17.2

$$\begin{aligned} Y'_{\text{S-video-625}} &= \frac{7}{10} Y'_{601} &&+ \frac{3}{10}(-sync) \\ C_{\text{S-video-625}} &= \frac{7}{10} C &&+ \frac{3}{10}\left(\frac{burst}{2}\right) \end{aligned}$$

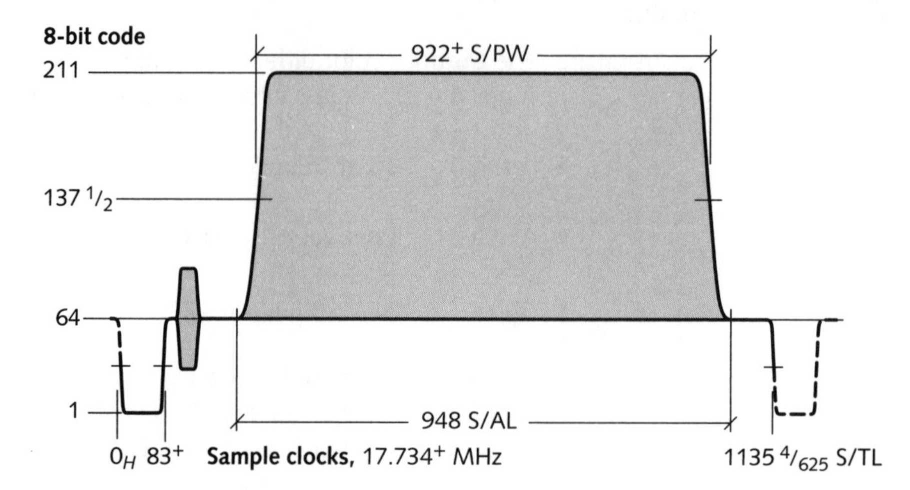

Figure 17.2 **625/50 composite digital waveform,** showing the range of the luma component. The chroma component will contribute to excursions outside this range.

Composite PAL encoding

A composite PAL signal is formed by summing luma (Y') and modulated chroma (C) signals, along with sync and colorburst.

The picture-to-sync ratio for PAL is 7:3, leading to a composite signal formed as follows, where *sync* is taken to be unity when asserted and zero otherwise, and *burst* is in the range ±1:

Eq 17.3

$$PAL = \frac{7}{10}\left(Y'_{601} + C\right) + \frac{3}{10}\left(\frac{burst}{2} - sync\right)$$

Composite digital interface, $4f_{SC}$

Digital composite PAL – also known as the 625/50 version of $4f_{SC}$ – is formed by scaling and offsetting Equation 17.3 so that in an 8-bit system blanking (zero in the equation) is at codeword 64 and reference white is at codeword 211. Figure 17.2 above shows a waveform drawing of luma in a 625/50 composite digital interface, with 7.5-percent setup.

The electrical and mechanical interface for 625/50, $4f_{SC}$, is derived from the Rec. 656 parallel and serial interfaces. See page 248.

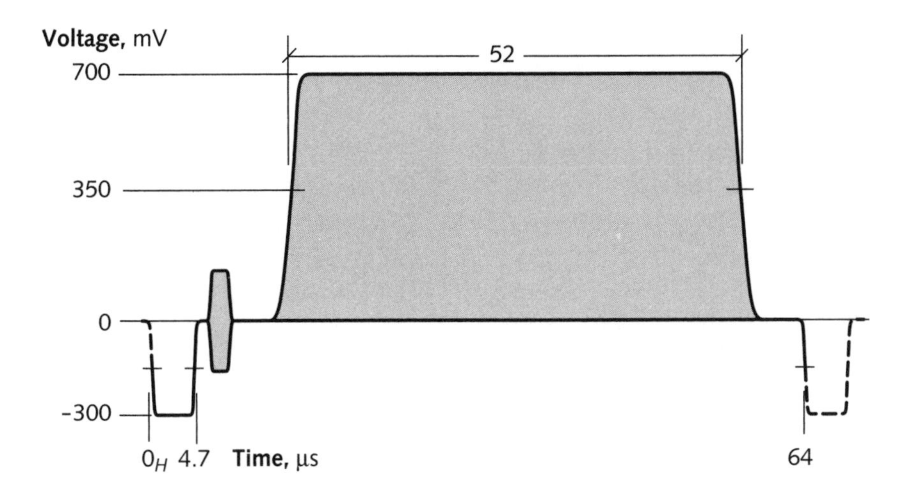

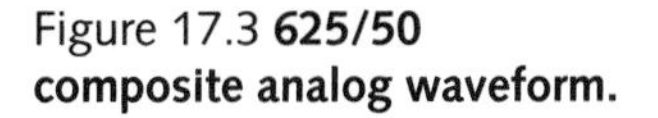
Figure 17.3 **625/50 composite analog waveform.**

Composite analog PAL interface

Figure 17.3 above shows a waveform drawing of luma in a 625/50 composite analog interface. The interface is detailed in *Analog electrical interface* on page 247. The excursion of an analog composite 625/50 PAL signal with zero chroma, from synctip to reference white, is 1 V_{PP}, comprising 700 mV of picture and 300 mV of sync.

Electrical and mechanical interfaces 18

Video interfaces use electrical and mechanical parameters from a small set of standards.

Analog electrical interface

Analog video is usually conveyed in the studio as a voltage on an unbalanced coaxial cable into a pure-resistive impedance of 75 Ω. Reference blanking level – zero in the equations used in chapters 12 through 17 – corresponds to a level of 0 V_{DC}. For component analog video, reference white – unity in the equations – often corresponds to 700 mV, except for signals with 10:4 picture-to-sync ratio, where white is $^5/_7$ V, or about 714 mV.

Analog mechanical interface

Figure 18.1
BNC connector.

It is standard for studio analog video to use a *BNC* connector that conforms to IEC 169-8. That standard defines a 50 Ω connector, but video systems use an impedance of 75 Ω, and video standards encourage the use of connectors whose impedance is 75 Ω. A set of three connectors is used for *R'G'B'* interface. In video, sync is usually inserted on the luma or green component. In computing, separate sync is preferred.

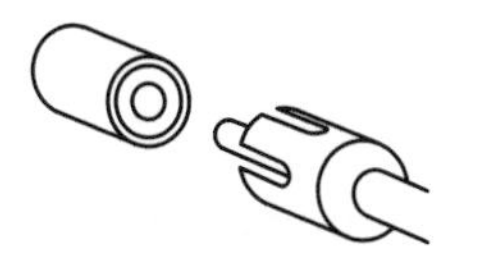

Figure 18.2
RCA phono connector.

Industrial and consumer equipment interfaces composite video at *baseband*, directly as a voltage from DC to about 5 MHz, using an *RCA phono* connector. An accompanying audio signal uses a separate phono connector; accompanying stereo audio uses a separate pair of phono connectors.

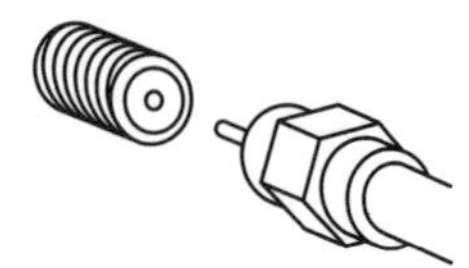

Figure 18.3
Type-F connector.

Consumer equipment often uses *radio frequency (RF) modulation*, where the composite video signal is modulated onto a VHF radio frequency carrier. In NTSC countries this is usually at the same frequency as either channel 3 or channel 4. This interface uses a *type-F* connector. Audio is modulated onto the standard audio subcarrier, and so is conveyed on the same cable.

In *S-video-525, Y′/C3.58*, on page 225, I describe the S-video interface, and sketch its connector.

Parallel digital interface

Recommendation ITU-R BT.656-1, *Interfaces for digital component video signals in 525-line and 625-line television systems operating at the 4:2:2 level of Recommendation ITU-R BT.601*. Geneva: ITU, 1990.

SMPTE 125M-1992, *Component Video Signal 4:2:2 – Bit-Parallel Digital Interface.*

Rec. ITU-R BT.656 defines a parallel electrical interface to convey signals coded according to Rec. 601. This interface has been adopted for 525/59.94 systems in SMPTE 125M-1992, and for 625/50 systems in EBU Tech. 3267. The interface carries eight (or optionally ten) data pairs, and a clock pair carrying a 27 MHz clock signal, at ECL levels on a DB25S connector.

Synchronization is achieved through a *timing reference sequence* (TRS) initiated by 8-bit codewords 0 and 255, which are protected from appearing elsewhere in the data stream. A TRS immediately precedes active video (*start active video*, SAV); another immediately follows active video (*end active video*, EAV).

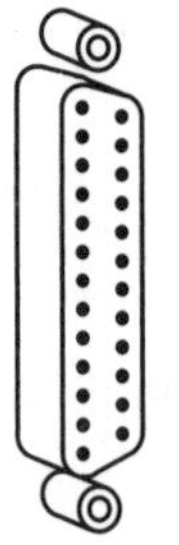

Figure 18.4 **Rec. 656 connector (DB25S).**

Rec. 656 defines a method of conveying ancillary data during intervals not occupied by TRS or by active video. Standards are emerging to carry digital audio in ancillary data.

An 8-bit transmitter places binary zero on the two low-order bits of the interface, except during the codeword-255 intervals of the TRS sequence, when it asserts binary ones. An 8-bit receiver ignores the two low-order bits of the interface.

SMPTE 244M-1993, *System M/NTSC Composite Video Signals – Bit-Parallel Digital Interface.*

SMPTE 267M, *Bit-Parallel Digital Interface – Component Video Signal 4:2:2 16×9 Aspect Ratio.*

The Rec. 656 parallel interface applies to component 4:2:2 digital video. SMPTE 244M standardizes a variant for $4f_{SC}$ NTSC; that interface carries a digitized representation of analog sync instead of TRS.

SMPTE 267M standardizes a version of the component 4:2:2 interface with 16:9 aspect ratio.

Serial digital interface

SMPTE 259M-1993, *10-Bit 4:2:2 Component and $4f_{SC}$ Composite Digital Signals – Serial Digital Interface.*

SMPTE 259M specifies a serial interface for 4:2:2 component digital video, and composite $4f_{SC}$ NTSC and PAL. The same 4:2:2 interface is adopted in Rec. 656. The serial composite $4f_{SC}$ interface uses TRS to achieve sync, rather than using digitized analog sync.

In the parallel interface, a dedicated clock signal accompanies the data. In the serial interface it is necessary for a receiver to recover the clock from the coded bitstream. The coded bitstream must therefore contain significant power at the coded bit rate. Also, the coded stream must contain little power at very low frequencies: The code must be *DC-free*. To enable economical equalizers, the ratio of the highest to lowest frequency components that are required for correct recovery of the signal should be as small as possible. A ratio of about 2:1 – where the coded signal is contained in one *octave* of bandwidth – is desirable. These considerations argue for a high clock rate. But it is obviously desirable to have a low clock rate, so the choice of a clock rate is a compromise between these demands.

A previous version of Rec. 656 specified *8b9b* coding: Each 8-bit word of the Rec. 601 stream was mapped through a lookup table to a 9-bit code, and that code was serialized. The scheme is now abandoned.

The serial interface uses scrambled coding, where the data stream is serialized, then passed through a shift register arrangement with exclusive-or taps implementing a characteristic function $x^9 + x^4 + x + 1$. Scrambling techniques using a single scrambler are well known. But the Rec. 656 scrambler has a second-stage scrambler, whose characteristic function is $x + 1$. The two cascaded stages offer improved performance over a conventional single-stage scrambler.

The data rate at the interface is the word rate times the number of bits per word. It is standard to serialize ten-bit data, even if the two LSBs are zero.

The bit rates for versions of the serial interface are these:

- For composite $4f_{SC}$ NTSC video, about 143 Mb/s
- For composite $4f_{SC}$ PAL video, about 177 Mb/s
- For 4:2:2 components according to Rec. 601, 270 Mb/s
- For variants of 4:2:2 sampled at 18 MHz to achieve 16:9 aspect ratio, 360 Mb/s

The serial interfaces use ECL levels, 75 Ω impedance, BNC connectors, and coaxial cable.

Fiber optic interfaces

Fiber optic interfaces for digital video are not yet standardized. You can expect fiber optic interfaces to be relatively straightforward adaptations of the serial versions of Rec. 656. Their introduction awaits the specification of appropriate components such as fiber optic cable and connectors.

Broadcast standards 19

Table 19.1 below summarizes the six major standards used for conventional broadcast television. This chapter will give details of each system.

The latest edition of ITU-R Report 624 – at the time of writing, Rep. 624-4, from 1990 – is the definitive source of detailed technical information regarding broadcast television standards. You may also wish to consult the booklet issued by Rohde & Schwarz.

CCIR and FCC tv standards. Munich: Rohde & Schwarz, 1989, PD 756.6981.21.

This chapter discusses basic parameter values, but not tolerancing issues. Numeric values in this chapter are exact unless indicated otherwise.

Table 19.1 **Summary of broadcast standards.** This table summarizes the scanning system, video bandwidth, color encoding technique, and subcarrier frequency of standard television broadcast systems. The predominant systems, 525/59.94 NTSC and 625/50 PAL, are emphasized by shading.

Video bandwidth (MHz)	4.2	4.2	5.0	5.5	5.5
525/59.94 scanning:					
ITU-R System	**M/NTSC**	**M/PAL**			
Colloquial	NTSC	PAL-M, PAL-525			
f_{SC}	3.579454545^{+}	3.575611889^{-}			
Where used	N.A., Japan	Brazil			
625/50 scanning:					
ITU-R System	**N/PAL**		**B,G,H/PAL**	**I/PAL**	**L,D,K/SECAM**
Colloquial	PAL-N, PAL-3.58		PAL	PAL	SECAM
f_{SC}	$3.582056025^{=}$		$4.433618750^{=}$	$4.433618750^{=}$	
Where used	Argentina		Europe, Australia	Britain	France, Russia

ITU-R, former CCIR

International agreement is necessary to avoid contention in the use of the RF spectrum. The *International Telecommunications Union* (ITU) is the body that achieves international agreement on matters concerning RF spectrum. Its branch, the ITU-R (formerly the International Radio Consultative Committee, or CCIR), is responsible for setting television broadcast standards. ITU-R agreements result in *de jure* or regulatory standards in its member states. Those standards effectively have the force of law.

The ITU jurisdiction over radio transmission has, over the years, extended into the domain of international program exchange on videotape. Because videotape recording standards have historically been intimately related to studio signals, ITU-R is involved in studio video standards. Its agreements in this domain are called *Recommendations* in the language of international standards; they amount to international voluntary standards similar to those of ISO or IEC.

The ITU-R also collects information from member states about technical aspects of broadcasting. The ITU-R publishes this information in *Reports*. ITU-R *Reports* are not standards, although they are important guides to implementors and users.

ITU-R scanning nomenclature

ITU-R notation for television transmission combines scanning lines per frame and field rate, as described in the previous section, with various baseband video and RF transmission parameters, into a *system* designation indicated by a single letter. ITU-R designations apply to the radiated signal: There is no standardized system of nomenclature for signals at baseband, or signals on videotape or disc.

System M refers to 525/59.94 scanning, with a video bandwidth of 4.2 MHz and a channel spacing of 6 MHz. There are many system designations for variants of 625/50 scanning, such as systems B, G, H, I.

Unless it is important to indicate bandwidth and RF properties, these should be referred to collectively as *625/50*. The headings in the tables in this chapter show the ITU-R nomenclature along with the informal nomenclature.

The ITU-R denotes a color video signal by its scanning system and bandwidth, denoted by a letter, as described in the previous section, followed by a slash and its color encoding method (NTSC, PAL, or SECAM). In informal nomenclature, a color video signal is denoted informally by its scanning standard (total line count and a frame rate) followed by its color standard; for example, 625/50/PAL/4.43.

M/NTSC (NTSC)

North America and Japan use system M/NTSC, which uses 525/59.94 scanning and NTSC color encoding with a subcarrier frequency of about 3.58 MHz.

M/PAL (PAL-M, PAL-525)

Brazil uses System M (525/59.94) scanning parameters and PAL color encoding with a subcarrier close to 3.58 MHz (but not exactly that of NTSC).

PAL-M counts lines in 625/50 style, with line 1 containing the first broad pulse coincident with line sync, with the count continuing throughout the entire frame. This leads to the somewhat embarrassing situation in Brazil that despite their rasters being identical, monochrome and color systems have their lines numbered differently.

N/PAL (PAL-N, PAL-3.58)

System N refers to 625/50 scanning with a bandwidth of about 4.2 MHz. Argentina uses system N/PAL with a subcarrier close to 3.58 MHz (but not exactly the same as NTSC). Monochrome ITU-R System N (monochrome PAL-N) also has a picture-to-sync ratio of 10:4. In Argentina, monochrome and color signals have the same frame and line frequencies, but their sync amplitude differ.

B,G,H,I/PAL (PAL)

Systems B, G, H, and I PAL all have 625/50 scanning and a video bandwidth ranging to 6 MHz. Australia and Europe (excluding France) use B, G, H/PAL; Britain uses I/PAL. The systems differ in their video bandwidths, channel spacing, and sound subcarrier frequencies. Table 19.2 opposite summarizes those parameters.

D,K/SECAM (SECAM)

Systems D and K refer to 625/50 scanning. SECAM uses FM modulation of the color subcarrier and transmits *U* and *V* signals line-sequentially. SECAM is used nowadays only as a transmission standard. Program production in SECAM countries is universally performed in PAL, then transcoded to SECAM prior to transmission. SECAM is used only in 625/50 systems. System L/SECAM is used in France; D/SECAM and K/SECAM are used in the nations of the former USSR. The versions differ in their video and audio modulation.

Summary of parameters

Table 19.2 opposite summarizes RF and audio characteristics of the major systems. Table 19.3 opposite summarizes the color encoding parameters of broadcast video standards.

Table 19.2 **RF and audio characteristics of broadcast standards.**

ITU-R system	*Informal*	*Video bandwidth (MHz)*	*Channel spacing (MHz)*	*Sound carrier (MHz)*	*Sound modulation*	*Where used*
M/NTSC	**NTSC**	4.2	6	4.5	FM, ±25 kHz	North America, Japan
M/PAL	**PAL-M (PAL-525)**	4.2	6	4.5	FM, ±25 kHz	Brazil
N/PAL	**PAL-N (PAL-3.58)**	4.2	6	4.5	FM, ±25 kHz	Argentina
B/PAL	**PAL**	5	7	5.5	FM, ±50 kHz	Australia
G,H/PAL	**PAL**	5	8	5.5	FM, ±50 kHz	Europe
I/PAL	**PAL**	5.5	8	6	FM, ±50 kHz	Britain
L/SECAM	**SECAM**	6 (+ve *video*)	8	6.5	AM	France
D,K/SECAM	**SECAM**	6	8	6.5	FM, ±50 kHz	Russia

Table 19.3 **Subcarrier characteristics of broadcast standards.**

ITU-R system	*Informal*	*Scanning*	*f_{SC}/f_H ratio*		*f_{SC} (MHz)*	
M/NTSC	**NTSC**	525/59.94	$\frac{455}{2}$	= 227.5	3.579 545 454$^+$	
M/PAL (Brazil)	**PAL-M (PAL-525)**	525/59.94	$\frac{909}{4}$	= 227.25	3.575 611 889$^-$	
N/PAL (Argentina)	**PAL-N (PAL-3.58)**	625/50	$\frac{917}{4}+\frac{1}{625}$	= 229.2516	3.582 056 250	(exact)
B,G,H,I/PAL	**PAL**	625/50	$\frac{1135}{4}+\frac{1}{625}$	= 283.7516	4.433 618 750	(exact)
D,K,L/SECAM	**SECAM**	625/50	272	(D_B)	4.250 000 000	(exact)
			282	(D_R)	4.406 250 000	(exact)

Test signals 20

This chapter summarizes the principal test signals used for video testing.

Colorbars

EIA 189-A-1976, *Encoded Color Bar Signal* (formerly known as EIA RS-189-A).

SMPTE EG 1-1990, *Alignment Color Bar Test Signal for Television Picture Monitors.*

Figure 20.1 below is a sketch of the image produced by the classic colorbar test signal. The upper two thirds of the image is formed from scan lines that result from a binary sequence of red, green, and blue values either zero or unity. The narrow, central region is similar, except that the sequence is reversed and green is set to zero. In the usual *75 percent bars*, white comprises a mixture of primary components at 100 percent, but the other bars are formed from primary components reduced in amplitude to 75 percent; this prevents the composite excursion from exceeding 100 units.

Figure 20.1 **Colorbars.**

R'
G'
B'

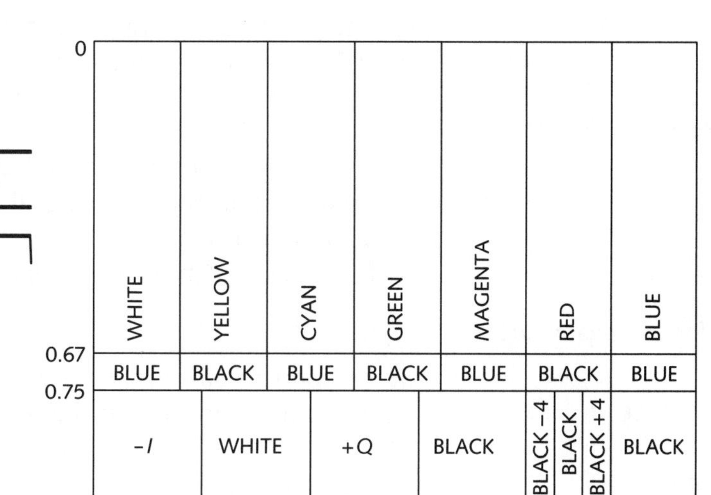

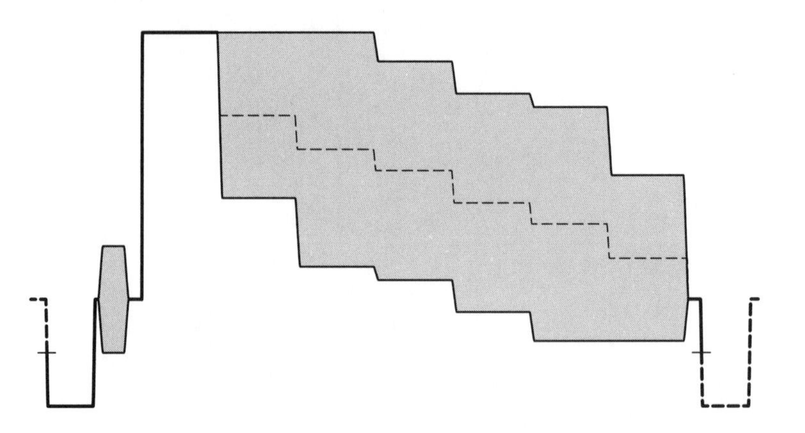

Figure 20.2 **Colorbar waveform.**

Figure 20.2 shows the NTSC composite waveform of a scan line in the upper region of 75 percent colorbars.

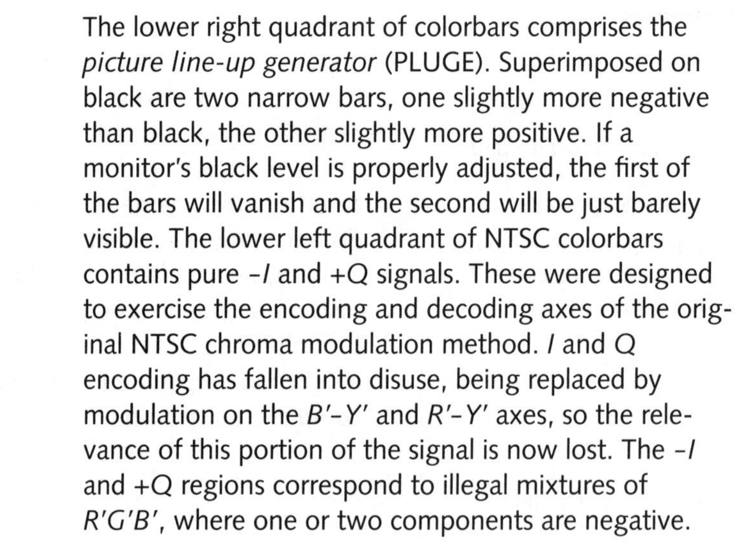

Figure 20.3 **Picture line-up generator (PLUGE).**

The lower right quadrant of colorbars comprises the *picture line-up generator* (PLUGE). Superimposed on black are two narrow bars, one slightly more negative than black, the other slightly more positive. If a monitor's black level is properly adjusted, the first of the bars will vanish and the second will be just barely visible. The lower left quadrant of NTSC colorbars contains pure -*I* and +*Q* signals. These were designed to exercise the encoding and decoding axes of the original NTSC chroma modulation method. *I* and *Q* encoding has fallen into disuse, being replaced by modulation on the *B'*-*Y'* and *R'*-*Y'* axes, so the relevance of this portion of the signal is now lost. The -*I* and +*Q* regions correspond to illegal mixtures of *R'G'B'*, where one or two components are negative.

Frequency response

Frequency response of a video system can be tested using a frequency sweep signal. In operational testing, a *multiburst* signal having several discrete frequencies is usually used instead. See Figure 20.12, on page 264.

Differential gain, DG

I introduced *Linearity* on page 16, and explained that a system is linear iff this condition is satisfied:

Eq 20.1

$$f(a+b) \equiv f(a) + f(b)$$

Figure 20.4 **Modulated ramp waveform.**

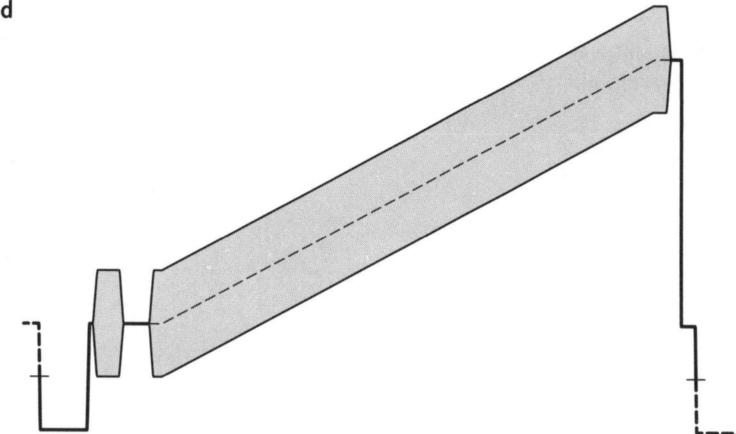

A system or subsystem with linear amplitude response has a transfer function whose output plots as a straight-line function of its input. You could characterize amplitude nonlinearity directly by the departure of the transfer function from an ideal straight line; ADC and DAC components and subsystems are specified this way. But in video, amplitude nonlinearity is usually characterized by a parameter *differential gain* (DG) that is easier to measure.

To measure DG, you present to a system the sum of a high-frequency signal and a low-frequency signal. It is standard to use, as the high-frequency signal, a sine wave at subcarrier frequency. The low-frequency signal is either a ramp or a staircase. Figure 20.4 above shows the preferred modulated ramp signal. The alternative modulated stair is shown in Figure 20.5 overleaf.

Ideally, when measuring DG, the high-frequency sine wave component should emerge at the output of the system having its amplitude independent of the low-frequency component upon which it is superimposed. The DG measuring instrument has a filter that rejects (discards) the low-frequency component; DG is then determined from the amplitude of the remaining high-frequency (subcarrier-based) component. Nonlinear amplitude response is revealed if the amplitude of the sine wave varies across the line.

Figure 20.5 **Modulated 5-step stair waveform.**

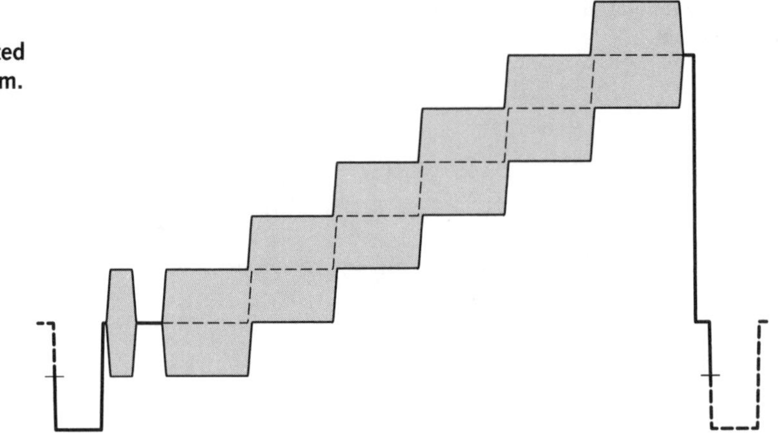

DG measurement directly applies the linearity principle in Equation 20.1. The system's input is the sum of two signals, $a + b$, where a is the high frequency signal and b is the ramp or staircase. The system's output is $f(a + b)$; we compare this to the ideal $f(a) + f(b)$.

Differential phase, DP

A potential defect in a video system is that phase may change as a function of DC (or low-frequency) amplitude. This error is known as *differential phase* or DP. Although in principle DP could be a problem at any frequency, in practice it is measured only at the subcarrier frequency. DP can be measured using the same test signal used to measure DG: a constant-amplitude sine wave at subcarrier frequency, superimposed on a ramp or staircase.

At the output of the system, low-frequency information is filtered out as in DG measurement, but this time the subcarrier-frequency signal is demodulated by a circuit identical to an NTSC chroma demodulator. The demodulated components can be displayed in vectorscope presentation, or the component at sine phase can be displayed in a waveform presentation.

Pulse signals

Pulse waveforms in video, such as sync, have waveforms chosen so as to be contained within the video bandwidth. If a pulse with a rapid transition is used, its high-frequency energy will not traverse the video system and is likely to produce ringing or distortion.

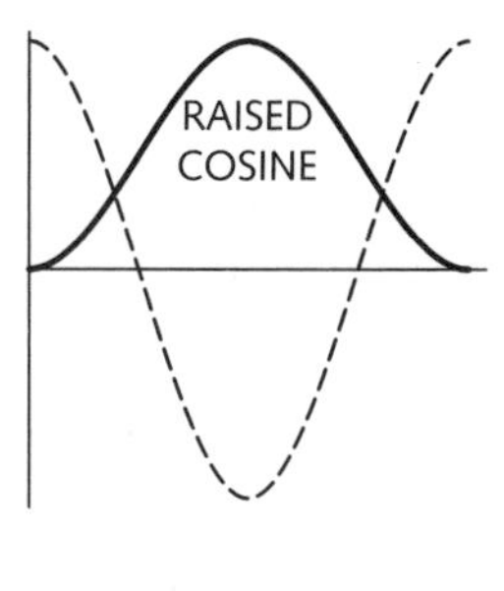

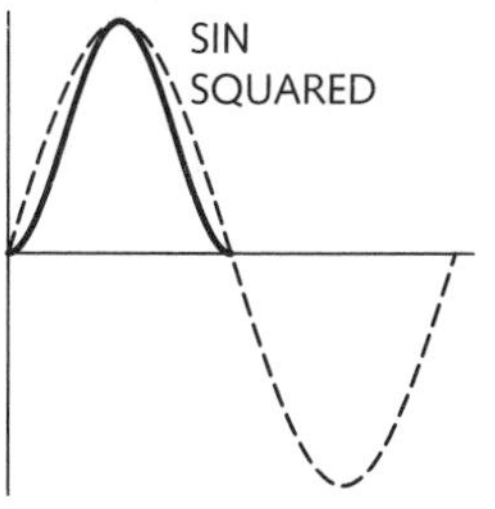

Figure 20.6 **Raised cosine, sine squared pulses.**

The most common pulse shape used is the *raised cosine*. This signal comprises a single (360°) cycle of a cosine wave that has been scaled by -0.5 and "raised" +0.5 units: It is a single pulse with a cosine shape. It is also known as *sine squared* (sin^2), because the pulse can be expressed as the square of a sine signal; a 180° segment of a sine wave, squared, forms a single pulse.

In the analog domain, a raised cosine pulse can be obtained by passing a very short impulse through a suitable filter. A raised cosine step, for high-quality sync, can be obtained by passing a step waveform through a similar filter.

Margaret Craig, *Television Measurements: NTSC Systems* Beaverton, OR: Tektronix, 1989, p/n 25W-7049.

A raised cosine pulse is often used for testing in video, because a single pulse can exercise a wide range of frequencies. Nonuniform frequency response is revealed by ringing in the system's response. Nonlinear phase response is revealed by a lack of symmetry. The amplitude of and phase of ringing is often characterized using a somewhat arbitrary measure called the *K factor*, whose definition is related to the subjective effect of the ringing. A system's K factor is related to the uniformity of its frequency response.

The duration of a raised cosine pulse, or any similar pulse with a single major lobe in its waveform, is measured in terms of its *half-amplitude duration*, HAD. The HAD of a raised cosine pulse used in testing is denoted by a parameter *T*. In 525/59.94 video, *T* is given the value 125 ns – a *T* pulse occupies a bandwidth of about 8 MHz. This is well beyond the bandwidth limit of legal studio video, but the pulse is useful for stress testing. A 2*T* pulse having HAD of 250 ns exercises a bandwidth of 4 MHz, just within the edge of legal broadcast video, so a 2*T* pulse should traverse a properly functioning video system.

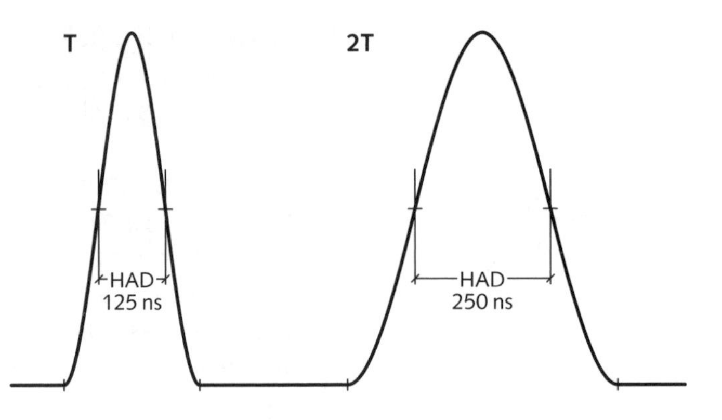

Figure 20.7 **T pulse waveforms.** *T* and 2*T* pulses are shown, in the time domain.

Figure 20.7 above shows the waveforms of *T* and 2*T* pulses. Figure 20.8 below shows their frequency spectra, along with the spectrum of a triangle-shaped pulse having the same rise time.

The rise time (from 10 percent to 90 percent) of a *T* step is not the same as the rise time of a *T* pulse.

Although the term *raised cosine* or *sine squared* properly applies to a complete pulse from zero through unity to zero, it is loosely applied to a transition from one level to another. For example, the leading edge of analog sync ideally has a raised cosine shape. A *T step* pulse is shown in Figure 20.9 opposite.

Modulated 12.5T, 20T pulses

In *Phase response (group delay)*, on page 56, I described how the time delay through a filter can change as a function of frequency. This phenomenon can be a problem in a video system, which can be considered to be a long cascade of filters. This condition is tested through use of a signal having two widely spaced frequency components.

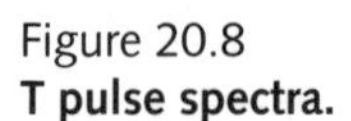
Figure 20.8
T pulse spectra.

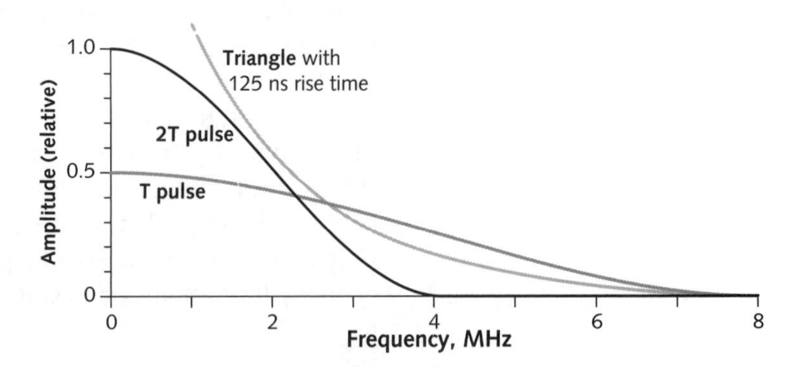

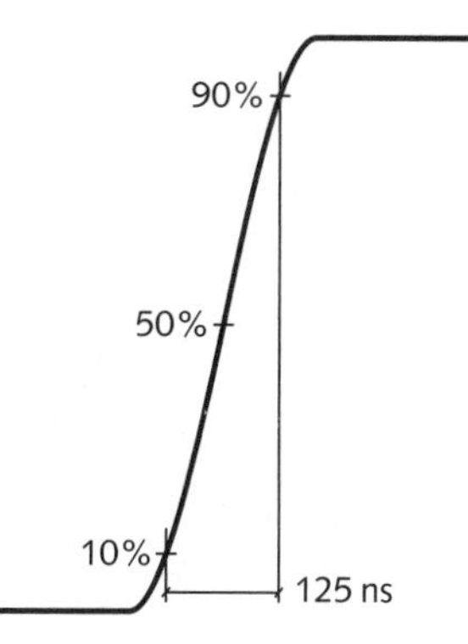

Figure 20.9 **T step pulse waveform.**

The modulated 12.5*T* pulse used to test group delay is shown in Figure 20.10 below. The signal comprises a low-frequency raised cosine of half amplitude duration 12.5*T* (1.5625 µs), upon which is superimposed a subcarrier-frequency cosine wave that has been amplitude-modulated by the low-frequency raised cosine. In an ideal system the resulting signal has an envelope that matches the raised cosine above and sits precisely at the baseline. Should the subcarrier frequency suffer delay, the envelope exhibits distortion; this is particularly evident along the baseline. Sometimes a 20*T* pulse is used instead of 12.5*T*. Note that *T* denotes the HAD of the *envelope*, not of the high frequency waves.

Figures 20.11 through 20.14 overleaf show several standard test signals that combine various elements discussed in this chapter.

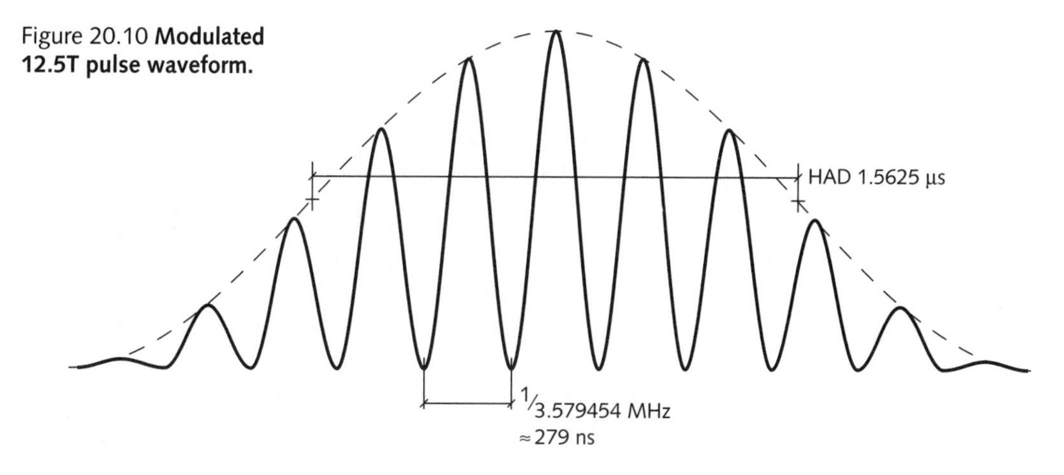

Figure 20.10 **Modulated 12.5T pulse waveform.**

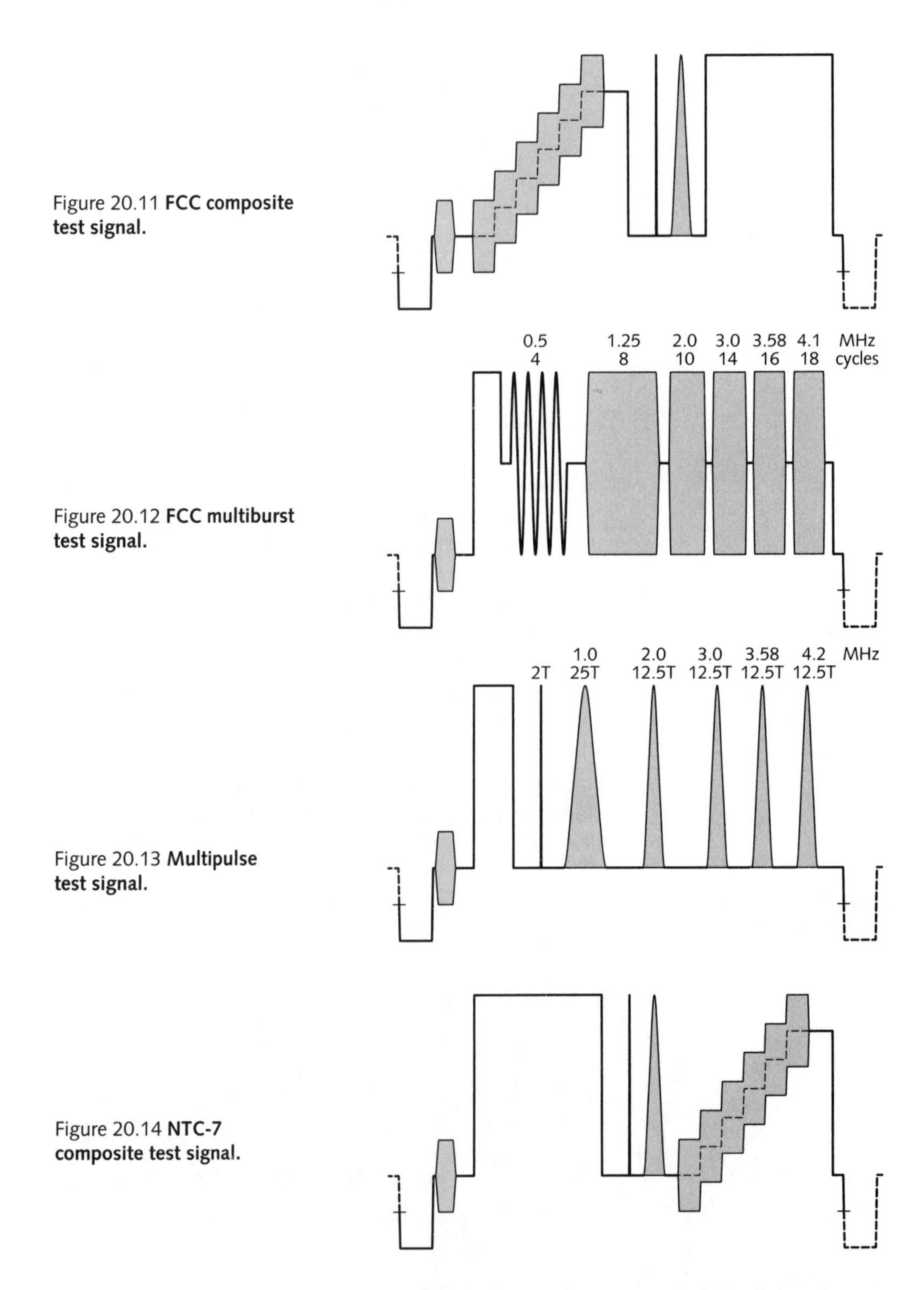

Figure 20.11 **FCC composite test signal.**

Figure 20.12 **FCC multiburst test signal.**

Figure 20.13 **Multipulse test signal.**

Figure 20.14 **NTC-7 composite test signal.**

Timecode 21

This chapter gives technical details concerning timecode, as used in video, film, audio recording, editing, and sequencing equipment.

Introduction

Timecode systems assign a number to each frame of video analogously to the way that film is manufactured with edge numbers to allow each frame to be uniquely identified. Time data is coded in binary coded decimal (BCD) digits in the form HH:MM:SS:FF, in the range 00:00:00:00 to 23:59:59:29 for 30 Hz frame rate systems. There are timecode variants for systems having 24, 25, 29.97, and 30 frames per second.

SMPTE RP 135-1990, *Use of Binary User Groups in Motion-Picture Time and Control Codes.*

In addition to the 32 bits required for eight-digit time data, timecode systems accommodate an additional 32 user bits per frame. User bits may convey one of several types of information: a second timecode stream, such as a timecode from an original recording; a stream of ASCII/ISO characters; motion picture production data, as specified in SMPTE RP 135; auxiliary BCD numerical information, such as tape reel number; or nonstandard information. A group of 4 user bits is referred to as a *binary group*. The information portion of timecode thus totals 64 bits per frame.

A number of synchronization bits are appended to the 64 information bits of timecode in order to convey timecode through a channel. Sixteen synchronization bits are appended to form 80-bit longitudinal timecode

(LTC). Eighteen sync bits and 8 CRC bits are appended to form 90-bit vertical interval timecode (VITC) that can be inserted into a video signal.

The BCD coding of time data has two implications. First, since no BCD digit can contain all ones, the all-ones code is available for other purposes. Second, the high-order bits of certain timecode digits are available for use as flags. These flag bits are described later.

The color frame flag is asserted when the least significant bit of the timecode frame number is intentionally locked to the color frame sequence of the associated video, that is, locked to Frame A and Frame B of SMPTE 170M. See *Two-frame sequence*, on page 222.

Dropframe timecode

In 25 Hz video, such as in 625/50 video systems, and in 24 Hz film, there is an exact integer number of frames in each second. In these systems, timecode has an exact correspondence with clock time. In 525/59.94 systems, there is not an integer number of frames per second. The remainder of this section deals with the *dropframe* mechanism that is used to compensate timecode to obtain a reasonable approximation of clock time. Dropframes are not required or permitted when operating at 24, 25, or 30 frames per second.

During the transition from monochrome to color television, certain interference constraints needed to be satisfied among the horizontal, sound, and color frequencies. These constraints were resolved by reducing the 60.00 Hz field rate of monochrome television by a factor of exactly $^{1000}/_{1001}$ to create the color NTSC field rate of about 59.94 Hz. This led to the dropframe timecode that is familiar to anyone that has been involved in videotape editing.

Counting frames at the NTSC frame rate of 29.97 Hz is slower than real time by the factor $^{1000}/_{1001}$, which, if left uncorrected, would result in an apparent cumulative error of about +3.6 seconds in an hour. To avoid this error, timecode systems are designed so that on average once every *1000* frames a frame number is dropped, that is, omitted from the counting sequence. Of course, it is only the timecode *number* that is dropped, not the whole frame!

Frame numbers must be dropped in pairs in order to maintain the relationship of timecode (even or odd frame number) to video color frame (A or B).

mm: ss: ff
x0:00:00
x1:06:20
x2:13:10
x3:20:00
x4:26:20
x5:33:10
x6:40:00
x7:46:20
x8:53:10

Dropping a pair of frames every 66 ⅔ seconds – that is, at an interval of one minute, six seconds, and twenty frames – produces the sequence in the table to the left. Although this sequence is not easily recognizable, it repeats after exactly ten minutes! This is a consequence of the ratios of the numbers: Two frames in 2000 accumulates 18 frames in 18 000, and there are 18 000 frames in 10 minutes (30 frames, times 60 seconds, times 10 minutes). To produce an easy-to-compute, easy-to-remember sequence of dropframe numbers, the key rule of dropframe timecode was devised:

> Drop frame numbers 00:00 and 00:01 at the start of every minute *except the tenth*.

In effect, a dropped pair that is due is delayed until the beginning of the next minute.

Longitudinal timecode, LTC

Timecode is recorded on studio videotape and audiotape recorders as a longitudinal track with characteristics similar or identical to those of an audio track. LTC is interfaced in the studio as an audio signal pair (0 dBm into 600 μW on a three-pin XLR connector).

Each frame time is divided into 80 bit cells; therefore the bit rate of timecode data is nominally 1.2 Kb/s. LTC is recorded using the *binary FM* technique, also known as *Manchester* code: Each bit cell has a transition at its start, a one bit has a transition in the middle of the cell and a zero bit does not. This coding is immune to the polarity reversals that sometimes occur in audio distribution equipment.

LTC is transmitted bit-serially, in four-bit *nibbles*; first a timecode nibble, then a user nibble, least significant bit first. This 64-bit stream is followed by a 16-bit sync word pattern that comprises the sequence 0, 0, twelve ones, 0, and 1. The sync pattern is distinguished from any data pattern, since the combination of BCD timecode digit coding and time/user digit interleaving inherently excludes any run of nine or more successive 1 bits. The sync pattern also identifies whether timecode is being read in the forward or reverse direction so that timecode can be recovered whether the tape is moving forward or backward.

At normal play speed, LTC can be decoded from tape as long as the playback system (heads, preamps) has an audio bandwidth out to about 2.4 kHz. To recover timecode at the shuttle rates of a high-quality studio VTR – about 60 times play speed – requires an audio bandwidth about 60 times higher. Due to the limitations of stationary head magnetic recording, longitudinal timecode from a VTR (or ATR) cannot be read at very slow speeds or with the tape stopped.

Vertical interval timecode, VITC

Vertical Interval timecode overcomes the disadvantage that LTC cannot be read with videotape stopped or moving slowly. With VITC, one or two video scan lines in the vertical interval of the video signal contain timecode data. VITC in 525/59.94 should be conveyed on line 14 (277). For videotape recording it is advisable to record VITC redundantly on two nonconsecutive lines, in case one line suffers a tape dropout. Lines 16 and 18 (279 and 281) are preferred. VITC identifies each field of video; a *field mark* bit is asserted for the even field (field 2).

For details concerning VITC line assignment, see SMPTE RP 164-1992, *Location of Vertical Interval Timecode.*

Each VITC line conveys 90 bits as 9 serialized bytes, each preceded by a 2-bit sync code (one, zero). The first 8 bytes contain the timecode information bits in LTC order.

The ninth byte contains a cyclic redundancy check (CRC) code that may be used for error detection (and possibly correction). CRC is computed as $G(x) = x^8 + 1$ across the 64 information bits and the 18 VITC sync bits. The CRC can be generated by an 8-bit shift register and an exclusive-or (XOR) gate. A CRC can be independently computed by the receiver from the information and sync bits; if the computed CRC does not match the transmitted CRC, then an error is known to have occurred.

The bit rate of VITC for 525/59.94 systems is one-half of the NTSC color subcarrier frequency, that is, one-half of $^{315}/_{88}$ MHz. A zero bit is at blanking level (0 IRE), and a one bit is at 80 IRE. The 0-to-1 transition of the first (start) bit is delayed 10.5 µs from the 50-percent point of the line sync datum (0_H). A decoder must use the sync bit transition at the start of the line to establish a decoder phase reference; it may or may not use the other sync transitions.

Editing

Timecode is fundamental to videotape editing. An edit is denoted by its in point (the timecode of the first frame to be recorded) and its out point (the timecode of the first frame beyond the recording). An edited tape can be described by the list of edits used to produce it. Each entry in such an edit decision list (EDL) contains the in and out points of the edited tape, and the in and out points of the source tape, along with tape reel number and/or other source and transition identification.

An edited tape is invariably recorded with continuous "nonbroken" timecode. Nearly all editing equipment treats the boundary between 23:59:59:29 and 00:00:00:00 as a timecode discontinuity; consequently, it is conventional to start the main program segment on tape with the code 01:00:00:00. If the tape includes the usual 1.5 minutes of bars and tone leader, then the tape will start near timecode 00:58:30:00.

Flag bits

Table 21.1 opposite illustrates the structure of timecode data. The flag bits have these meanings:

Dropframe flag	Asserted for dropframe timecode mode in 525/59.94 systems only.
Colorframe flag	Asserted when timecode is locked to the colorframe sequence of the associated video.
Parity (LTC only)	Computed such that the complete 80-bit LTC timecode contains an even number of zero bits (a.k.a. biphase mark phase correction).
Field Mark (VITC only)	Asserted for the even field.

Binary Group A	*Binary Group B*	*User-bit interpretation*
0	0	*Unspecified characters or data*
0	1	*ISO 646/2202 eight-bit characters*
1	0	*SMPTE RP 135-1990 data*
1	1	*Reserved/unassigned*

Further reading

SMPTE 12M-1990, *Time and Control Code – Video and Audio Tape for 525-Line/ 60-Field Television Systems.*

Timecode for 525/59.94 television is standardized in ANSI/SMPTE 12M. Timecode for 625/50 is standardized by the European Broadcasting Union document EBU Tech. 3097-E (1980).

SMPTE RP 136-1990, *Time and Control Codes for 24, 25 or 30 Frame-Per-Second Motion-Picture Systems.*

SMPTE RP 136 is a standard for magnetic recording of timecode on motion picture film at 24, 25, 29.97, or 30 frames per second. SMPTE RP 135-1990 standardizes a method of structuring user-bit data.

Table 21.1 **Timecode bit assignment table.** Transmission order is right to left, top to bottom – that is, least significant to most significant. For VITC, start with the rightmost column *VITC sync* and include the final row of *VITC sync* and *CRC* bits. For LTC, exclude the VITC sync columns and include two final rows of *sync A* and *sync B* bits.

	7	6	5	4	3	2	1	⇐0	⇐ VITC sync	
0	1st binary group				Frames units 0–9				0	1
8	2nd binary group *(or character 0)*				Color-frame flag	Drop-frame flag	Frame tens 0–2		0	1
16	3rd binary group				Seconds units 0–9				0	1
24	4th binary group *(or character 1)*				Parity/ field mark	Seconds tens 0–5			0	1
32	5th binary group				Minutes units 0–9				0	1
40	6th binary group *(or character 2)*				Binary group flag B	Minutes tens 0–5			0	1
48	7th binary group				Hours units 0–9				0	1
56	8th binary group *(or character 3)*				Binary group flag A	Unas-signed	Hours tens 0–2		0	1

In vertical interval code, the VITC sync and information bits are followed by 2 VITC sync bits, then 8 cyclic redundancy check (CRC) bits computed across the preceding 82 bits:

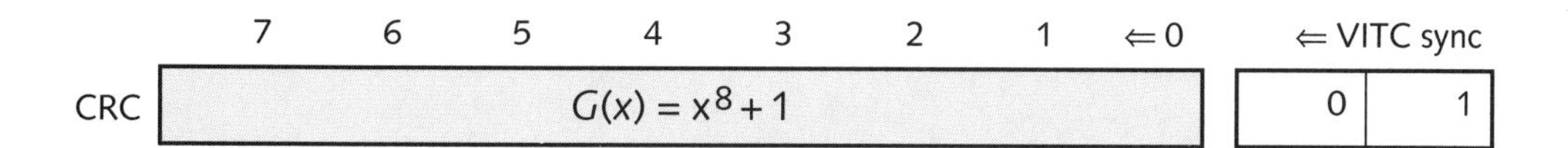

	7	6	5	4	3	2	1	⇐0	⇐ VITC sync	
CRC	$G(x) = x^8 + 1$								0	1

In LTC the 64 information bits are followed by 16 LTC synchronization bits:

	7	6	5	4	3	2	1	⇐0
sync A	1	1	1	1	1	1	0	0
sync B	1	0	1	1	1	1	1	1

Glossary of video signal terms

A

0_H datum — The reference point of horizontal (line) sync. The 50 percent point of the leading edge of the transition to synctip level.

0_V datum — The reference point of vertical (field) sync. The line sync – or in an interlaced system, alternately the halfway point between two line syncs – at the start of the first broad pulse of a field.

2-2 pulldown — The process by which motion picture film is transferred to interlaced video by scanning each film frame twice, once to produce the odd field and once again to produce the even field. This technique is commonly used to transfer motion picture film, running 4 percent faster than 24 frames per second, to 625/50 video. See also *3-2 pulldown*.

2-3 pulldown — See *3-2 pulldown*.

2:1 interlace — See *Interlace.*

+25 Hz offset — In a 625/50 PAL system, the color subcarrier frequency is an integer multiple of one-quarter the line rate, plus an *offset* of 25 Hz.

3-2 pulldown — The process by which motion picture film, running 0.1 percent slower than 24 frames per second, is transferred to 525/59.94 video. Alternate film frames are scanned first three then two times to form successive video fields.

3.58 MHz

More precisely, 3.579454 MHz, or, expressed exactly, $5 \times {}^{63}/_{88}$. The color subcarrier frequency of 525/59.94 NTSC video.

4.43 MHz

Expressed exactly, 4.433618750 MHz. The color subcarrier frequency of 625/50 PAL video.

4:1:1

A component digital system in which C_B and C_R chroma components are horizontally subsampled by a factor of 4, with respect to luma.

4:2:0

This confusing notation is sometimes used to designate a component digital system that subsamples C_B and C_R chroma components both vertically and horizontally by a factor of 2, with respect to luma.

4:2:2

A component digital system, as in Rec. 601, that represents a luma component Y', and two chroma components C_B and C_R that are horizontally subsampled by a factor of 2 with respect to luma. The corresponding 19 mm videotape format is known as *D-1*. There are 525/59.94 and 625/50 versions of 4:2:2.

4:4:4

A component digital system in which $R'G'B'$ or $Y'C_BC_R$ components are conveyed with equal data rate.

4:4:4:4

A 4:4:4 system (see above), augmented by a transparency or *alpha* channel sampled at the same rate as the luma component.

$4f_{SC}$

A composite digital system that utilizes a sampling frequency of 4 times the color subcarrier frequency. There are 525/59.94 and 625/50 versions of $4f_{SC}$. The corresponding 19 mm videotape format is known as *D-2*.

525/59.94/2:1

A scanning standard used primarily in North America and Japan, having 525 total lines per frame (of which approximately 483 contain picture information), a field rate of 59.94 Hz, and interlace. Without the "/2:1" notation, interlace is implicit. Known internationally as *ITU-R System M*. Color in 525/59.94 systems is conveyed in the studio using $R'G'B'$, $Y'C_BC_R$, or $Y'P_BP_R$ components and encoded for transmission using composite NTSC. Often loosely referred to as *525/60*. Often incorrectly called *NTSC*, which properly refers to a color standard, not a scanning standard. A raster notation such as 525/59.94 does not specify color coding.

625/50/2:1

A scanning standard used primarily in Europe and Asia, having 625 total lines per frame (of which 576 contain picture information), a field rate of 50 Hz, and interlace. Without the "/2:1" notation, interlace is implicit. A raster notation such as 625/50 does not specify color coding; color in 625/50 systems is commonly conveyed in the studio using $R'G'B'$, $Y'C_BC_R$, or $Y'P_BP_R$ components, and distributed by satellite using the MAC system, or distributed by terrestrial VHF/UHF television using composite PAL (although France, Russia, and certain other countries use SECAM). Often incorrectly called *PAL* or *SECAM*, which properly refer to color standards. Often incorrectly called *CCIR*.

Alpha

A component of a pixel indicating transparency – conventionally between zero (opaque) and unity (fully transparent – of the pixel's color components. Alpha can be *associated*, in which case the color component values have been premultiplied by the value of alpha; or *unassociated*, in which they have not been premultiplied. See also *Key*, on page 286.

$B'-Y'$, $R'-Y'$

A pair of color difference components, blue minus luma and red minus luma. The red, green, and blue primaries are gamma corrected prior to the formation of Y', $B'-Y'$ and $R'-Y'$. These color differences may subsequently be scaled to form U and V for subsequent composite encoding, scaled to form P_B and P_R for component analog transmission, or scaled to form C_B and C_R for component digital transmission.

Back porch

The time interval between the trailing edge of a normal line sync pulse and the start of picture information on the associated video line.

Black level

The level of a video signal that corresponds to black. Conventionally 7.5 IRE for System M and the archaic RS-343-A, and zero in other systems. See also *Pedestal*, on page 291.

Black-to-white excursion

The excursion from reference black to reference white. Conventionally 92.5 IRE ($^{37}/_{56}$ V, approximately 660 mV) for System M and the archaic RS-343-A, 700 mV in other analog systems, and codes 16 through 235 in component digital systems.

Blanking

1. The process of turning off the beam in a CRT, such as a vidicon camera tube or a display tube, in order for the tube to accomplish beam retrace without disturbing the picture.

2. The time interval – in the vertical domain, the horizontal domain, or both – during which a video signal is mandated to be at blanking level or sync level.

Blanking level

The level of the video signal during a blanking interval. 0 IRE by definition.

Broad pulse

A pulse, part of the field sync sequence, that remains at sync level for substantially longer than normal line sync and indicates vertical sync. It is standard for a broad pulse to have a duration of half a line time less the duration of normal line sync.

Burst

A brief sample of eight to ten cycles of color subcarrier inserted by an encoder onto the back porch of a composite color signal. Burst is inserted by an encoder to enable a decoder to regenerate the color subcarrier.

C_B, C_R

Versions of color difference components $B'-Y'$ and $R'-Y'$, scaled and offset for digital component transmission as standardized by Rec. 601 to encompass digital codes 16 through 240. See also $B'-Y'$, $R'-Y'$; P_B, P_R; and U, V.

In systems such as 525/59.94 and 625/50 using Rec. 601 luma, it is standard to use these scale factors:

$$C_B = 112 \cdot \frac{0.5}{0.886}; \qquad C_R = 112 \cdot \frac{0.5}{0.701}$$

CCIR

1. *Comité Consultatif Internationale des Radiocommunications*, International Radio Consultative Committee. A treaty organization, now renamed ITU-R.

2. Sometimes used incorrectly to denote 625/50/2:1 scanning.

CCIR Recommendation 601

Obsolete designation, now properly referred to as Recommendation ITU-R BT.601, or, colloquially, *Rec. 601*. See *Rec. 601*, on page 292.

CCIR Recommendation 709

Obsolete designation, now properly referred to as Recommendation ITU-R BT.709, or, colloquially, *Rec. 709*. See *Rec. 709*, on page 292.

CCIR System

1. Former international designation, now properly denoted *ITU-R System*. A set of parameters describing a television system, including scanning, video bandwidth, and RF channel spacing, as described in ITU-R Report 624.

2. Sometimes used incorrectly to denote 625/50/2:1 scanning.

CCIR System M

Former international designation, now properly denoted *ITU-R System M* or just *System M*. See *System M*, on page 296.

Chroma, *C*

1. A signal or set of signals that conveys color independent of luma or luminance.

2. Color conveyed as a pair of color difference signals such as $B'-Y'$, $R'-Y'$; C_B, C_R; P_B, P_R; I, Q; or U, V.

3. Two color difference components, *UV* or *IQ*, modulated onto a *subcarrier* using one of the NTSC, PAL, or SECAM composite color encoding techniques.

Chromaticity

1. An absolute specification of color in terms of *xy*, u^*v^*, or a^*b^* coordinates, as defined by the CIE.

2. Loosely, the chromaticity (1) of the red, green, and blue primaries and the chromaticity of the white reference of a television system.

Chrominance

1. Formally, the color of a scene or scene element independent of its luminance; usually, expressed in the form of CIE (x, y) chromaticity.

2. Loosely, *Chroma*; see above.

CIE

Commission Internationale de L'Éclairage, International Commission on Illumination. The international voluntary standards organization that sets colorimetry standards.

CIE D_{65}

The standard spectrum or chromaticity of white, representative of northern daylight and having a color temperature of approximately 6504 K. See *Reference white*, on page 292.

CIE luminance, CIE *Y*

See *Luminance*. The *CIE* notation is sometimes used to emphasize that the associated quantity is representative of tristimulus (linear-light) intensity, not a nonlinear, gamma-corrected quantity.

Coherent

Two or more periodic signals that are phase-locked to a common reference frequency are *coherent*. The color subcarrier of a studio-quality composite video signal is coherent with its sync.

Color

This term is used loosely in common speech but has a rather specific meaning in television. See *Trichromaticity* on page 297.

Colorburst

See *Burst*, on page 276.

Color difference

1. A numerical measure of the perceptual distance between two colors.

2. A signal that vanishes – that is, becomes identically zero – for pure luma without color. It is common in video to utilize a set of three signals, a luma signal and a pair of color difference signals, to convey a color image. The advantage of color-difference coding over $R'G'B'$ is that spatial and/or temporal filtering may be applied to reduce the information rate of the color difference components, with respect to luma, without being visually perceptible. Examples of color-difference pairs are $B'-Y'$, $R'-Y'$; C_B, C_R; P_B, P_R; I, Q; or U, V.

Color standard

The parameters associated with encoding of color information – for example, $R'G'B'$, $Y'C_BC_R$, or MAC *component* color standards, or NTSC, PAL, or SECAM *composite* color standards. Distinct from *Scanning standard*, on page 294.

Color subcarrier

A continuous sine wave signal used as the basis for quadrature modulation or demodulation of chroma in an NTSC or PAL composite color system. See also *Burst*, on page 276.

Color-under

A degenerate form of composite color in which the subcarrier is crystal-stable but not coherent with line rate. The term derives from the recording technique used in U-Matic (3⁄4-inch), Betamax, VHS, and 8 mm videotape recorders, where chroma is *heterodyned* onto a subcarrier whose frequency is a small fraction of that of NTSC or PAL. The heterodyning process loses the phase relationship of color subcarrier to sync.

Comb filter

A composite color decoder that incorporates one or more line delay elements in order to exploit the line-interleaving of modulated chroma (2) to separate chroma from luma. A comb filter provides better color separation than a notch filter, due to its suppression of *cross-color* and *cross-luma* artifacts.

Component

1. *Component color*; see below.

2. One of the set of three components or signals necessary to completely specify a color.

3. A channel or signal, such as transparency, that is spatially and temporally associated with an image, but is not itself representative of luminance, luma, or color information.

Component color

A video system that conveys three color component signals independently, free from mutual interference. Examples are $R'G'B'$, $Y'C_BC_R$, and MAC.

Composite color

A video system in which three color components are simultaneously present in a single signal. Examples are NTSC and PAL, which use the *frequency interleaving* principle to *encode* (combine) luma and chroma.

Composite sync

A deprecated term meaning *sync*. The word *sync* alone implies both horizontal and vertical elements, so *composite* is redundant. The adjective *composite* more meaningfully applies to *video* or *color*, so its use with *sync* is confusing.

Composite video

1. A video signal including a sync component.

2. A video signal including sync, luma, chroma, and burst components.

Constant luminance

In a color television system that dedicates one component to brightness information, the property that true (CIE) luminance reproduced at the display is unaffected by the values of the other two components. Conventional television systems, such as NTSC, PAL, and $Y'C_BC_R$, are close to exhibiting the constant luminance property, but because luma is computed as a weighted sum of gamma-corrected primaries, a certain amount of true luminance "leaks" into the color difference components as a second-order effect.

Cross-color

An artifact of composite color encoding and/or decoding that involves the erroneous interpretation of luma information as color. The cross-color artifact appears frequently when luma information having a frequency near that of the color subcarrier appears as a swirling color rainbow pattern.

Cross-luma — An artifact of composite color encoding and/or decoding that involves the erroneous interpretation of color information as luma. Cross-luma frequently appears as *dot crawl* or *hanging dots*.

CVBS — *Color, Video, Blanking*, and *Sync*. A European term for *composite video* (2).

D-1 — A *component* digital videotape format utilizing 19 mm tape cassettes to record component digital 4:2:2 $Y'C_BC_R$ video signals using either 525/59.94 or 625/50 scanning.

D-2 — A *composite* digital videotape format utilizing 19 mm tape cassettes and recording either 525/59.94 NTSC video or 625/50 PAL video, sampled at $4f_{SC}$. The D-2 recording formats are unrelated to the *D2-MAC* transmission system for 625/50; see below.

D2-MAC — A MAC system, based on 625/50/2:1 scanning, used in Europe for satellite broadcasting. D2-MAC is unrelated to *D-2* video recording. See D-2, above.

D_{65} — See *Reference white*, on page 292; and *CIE* D_{65}, on page 278.

Datum — See 0_H *datum* and 0_V *datum*, on page 273.

DC — Direct Current. Historically, an electrical current having no periodic – *alternating current*, or *AC* – component. Nowadays, having zero frequency.

Decoder — A circuit that takes composite color video such as NTSC or PAL, performs luma/chroma separation and chroma demodulation, and produces component video output such as $R'G'B'$.

Dot crawl	A cross-luma artifact that results from a notch filter decoder, appearing as fine luma detail crawling up a vertical edge in a picture that contains a saturated color transition.
Drive	A periodic pulse signal that conveys synchronization information. See *Vertical drive, VD*, on page 298; and *Horizontal drive, HD*, on page 285.
Dropframe	A timecode stream associated with 525/59.94 scanning, in which timecodes of the form *hh*:*tu*:00:00 and *hh*:*tu*:00:01 are omitted from the count sequence whenever *u* (the units digit of minutes) is nonzero, in order that counting frames obtains a close approximation to clock time. In applications where clock-time correspondence is important, this adjustment compensates for the field rate of 525/59.94 scanning being a factor of exactly $^{1000}/_{1001}$ slower than 60 Hz.
DVTR	Digital videotape recorder. See *VTR*, on page 298.
EIA	The U.S. Electronic Industries Association.
EIA No. 1	*EIA Industrial Electronics Tentative Standard No. 1* dated November 7, 1977, served to define the 525/59.94/2:1 NTSC system until the adoption of SMPTE 170M in 1994. See also *RS-170-A*, on page 294.
EIA RS-170, RS-170-A	See *RS-170* and *RS-170-A*, on page 294.
EIA RS-343-A	See *RS-343-A*, on page 294.
Encoder	A circuit that takes component video input (*e.g.*, *R'G'B'*), performs chroma modulation and luma/chroma summation, and produces composite color video (*e.g.*, NTSC or PAL).
Equalization	The correction of undesired frequency or phase response. Coaxial cable introduces a high-frequency rolloff that is proportional to cable length and to $^{1}/_{\sqrt{f}}$ (pronounced *one over root f*); this is corrected by

a subsystem called an *equalizer.* A naively designed analog lowpass filter, or a simple digital IIR filter, has poor phase response. This can be corrected by an *equalizer* filter section.

Equalization pulse

A sync pulse, part of field sync, that is approximately half the duration of a normal sync and occurs at 0_H or halfway between two 0_H data. The original purpose of equalization pulses was to eliminate the line pairing that would otherwise occur with cheap, passive sync separator circuits.

Even field

In a 2:1 interlaced system, the field that begins with a broad pulse starting halfway between two line syncs.

Excursion

The amplitude difference between two levels.

Exponential function

An arithmetic function of the form $y = a^x$, where a is constant. Exponential functions are rarely used in television. *Gamma correction* is a *power function,* and not an exponential function as is sometimes claimed.

Field

The smallest time interval of a video signal that contains a set of scanning lines covering the entire picture height, along with the associated sync elements. In a system with *noninterlaced* (or *progressive*) scanning, fields and frames are identical. In a system with *2:1 interlace*, a frame comprises two fields (*odd* and *even*); each contains half the scanning lines and half the picture lines of the frame. In historical analog usage of the term, each field in an interlaced system was considered to have a duration of an *odd* number of *half* lines (*e.g.*, 525 halflines, or 262 ½ lines). In modern digital usage of the term, the odd and even fields have durations of an integral number of lines that differ by one. In 525/59.94, field one has 263 lines and field two has 262 lines. See also *Interlace*, on page 285; *Even field*, on page 283; and *Odd field*, on page 290.

Field sync

The sync pulse pattern that occurs once per field and defines the start of a field. Field sync contains the 0_V

datum. In interlaced systems field sync is a sequence comprising pre-equalization pulses, broad pulses, and post-equalization pulses. In 525/59.94/2:1 systems there are six of each; in 625/50/2:1 systems there are five.

Frame

The time interval of a video signal that contains exactly one integral picture, complete with all of its associated sync elements. Typically measured between 0_V data.

Front porch

The time interval between the right-hand edge of the picture width and the immediately following 0_H datum.

Gamma

The numerical value of the exponent to which a video signal is raised – by a power function of the form v^{γ} – to obtain linear-light. A power function is inherent in the electron gun of a CRT display, so a gamma value characterizes a display system. The value is important throughout a television system because the inverse power function is applied at the camera, and R', G', B', and Y' signals are conveyed in *gamma-corrected* form throughout the system. Gamma in video typically has the numerical value $1/0.45$. Gamma properly has a value greater than unity, though the term is sometimes used loosely to refer to the reciprocal of true gamma.

Gamma corrected

A signal to which *gamma correction* has been applied – that is, a linear-light signal, such as a tristimulus value, raised to a power in the range $1/2$ to $1/3$. See *Gamma* above. Because gamma correction produces a signal that has similar response to the contrast sensitivity function of human vision, a video signal exhibits good perceptual uniformity: Noise or quantization error introduced into a video signal is approximately equally perceptible across the tone range of the system from black to white. Gamma correction also compensates for the nonlinear voltage-to-intensity transfer function inherent in a CRT display.

Gamma correction

The process by which a quantity representative of linear-light, such as CIE luminance or some other tristimulus signal, is transformed into a video signal by a power function with an exponent in the range ½ to ⅓. See *Gamma* above. Gamma correction is usually performed at a television camera or its control unit.

Hanging dots
A cross-luma artifact appearing as a fine stationary pattern of dark and light dots along a horizontal edge in a picture having a saturated vertical color transition, when a decoded by a comb filter. Hanging dots are particularly evident in the a colorbar test signal.

Horizontal blanking
The time interval – usually expressed in microseconds or sample counts, or sometimes as a fraction of line time – between the end of the picture information on one line and the start of the picture information on the following picture line.

Horizontal drive, HD
A pulse containing horizontal synchronization information that begins at the end of picture information on a line and ends at the trailing edge of sync.

I, *Q*
In-phase and Quadrature color difference components of NTSC; *U* and *V* rotated +33° and then axis-exchanged. NTSC chroma was originally based on *I* and *Q* color differences, where *Q* was bandwidth-limited more severely than *I*. Nowadays, NTSC color modulation is usually performed on equiband *U* and *V* components. Except for a possible bandwidth difference, it is impossible to tell from a composite analog NTSC signal whether it was encoded along *U*, *V*; *I*, *Q*; or any other pair of axes.

Interlace
A scanning standard in which alternate raster lines of a *frame* are displaced vertically by half the scan line pitch and displaced temporally by half the frame time to form an *odd field* and an *even field*. Also called 2:1 interlace. Examples are 525/59.94/2:1, 625/50/2:1, 1125/60.00/2:1. Systems with *high-order interlace* have been proposed but none has been introduced in practice, so modern usage of the term *interlace* implies 2:1 interlace. See also *Field*, on page 283.

IRE, IRE unit
: One-hundredth of the excursion from blanking level to reference white level. Originally standardized by the *Institute of Radio Engineers*, the predecessor of the IEEE. In ITU-R System M and the archaic EIA RS-343-A, where picture-to-sync ratio is 10:4, one IRE unit corresponds to $7\,^1\!/_7$ mV. In systems having picture to sync ratio of 7:3, one IRE unit corresponds to exactly 7 mV, although in those systems levels are usually expressed in millivolts and not IRE units.

ITU-R
: *International Telecommunications Union, Radiocommunications Sector.* Successor to the *Comité Consultatif Internationale des Radiocommunications* (International Radio Consultative Committee). A treaty organization that obtains international agreement on standards for radio and television broadcasting. The ITU-R BT series of Recommendations and Reports deals with television. Although studio standards do not involve broadcasting in a strict sense, they are used in the international exchange of programs, so they are under the jurisdiction of ITU-R.

ITU-R Recommendation BT. 601 Colloquially, *Rec. 601*. See Rec. 601.

ITU-R Recommendation BT. 709 Colloquially, *Rec. 709*. See Rec. 709.

Key
: A component signal indicating transparency, usually between zero (opaque) and unity (fully transparent), of the associated color components. A key is implicitly assumed not to have been premultiplied by the value of alpha, so a subsequent compositing operation is performed as: $composite = key \cdot FG + (1 - key) \cdot BG$. See also *Alpha*, on page 275.

Level
: The amplitude of a video signal, or one of its components, expressed in volts, millivolts, IRE units, or digital code value.

Line
: **1.** Scan line (horizontal). The term *line frequency* should be used with care in television because it may refer to *horizontal* scanning (as in *scan line*) or AC

mains frequency (which is usually similar to the *vertical* scan frequency).

2. See *TV line, TVL*, on page 297.

Line frequency

1. In a video raster, the frequency of horizontal scanning, about 15.75 kHz for conventional video.

2. AC power line (mains) frequency, typically 50 Hz or 60 Hz.

Line-locked

A digital video system having an integral number of samples per line. A line-locked system has coherent sampling and line frequencies, as in Rec. 601 or $4f_{SC}$ NTSC. Due to the *25 Hz offset* of the PAL subcarrier, $4f_{SC}$ PAL is not line-locked.

Line sync

The sync signal pulse transition that defines the start of a scan line. Line sync may be the start of a normal sync or the start of an equalization or broad pulse. See also 0_H *datum*.

Line time

The time interval between the 0_H datum of one line and the 0_H datum of the next.

Luma, Luma coefficients

A video signal representative of the monochrome – or roughly, lightness – component of a scene. In conventional 525-line and 625-line television systems, it is standard to compute luma from gamma-corrected red, green, and blue primaries as:
$Y'_{601} = 0.299\,R' + 0.587\,G' + 0.114\,B'$.

Luminance

Much confusion surrounds this term. In color science and physics, *luminance* is proportional to intensity (linear-light), and has the symbol Y. In television, the quantity loosely called *luminance* is a weighted sum of nonlinear (gamma-corrected) tristimulus components, and is properly denoted Y'. The term *luma* has been introduced to the television domain to refer to the nonlinear quantity in a way that avoids confusion. However, the term *luminance* is used loosely, and the prime on the symbol is often omitted. In practice, both the term *luminance* and the symbol Y are ambiguous: Whether the associated quantity is linear or nonlinear must be determined from context.

1. Formally, CIE Y tristimulus value. The integral of the spectral radiance of a scene, weighted by the luminous efficiency function of the CIE Standard Observer. The preferred unit for luminance is $cd \cdot m^{-2}$, or *nit*. Luminance is closely related to the brightness sensation of human vision.

2. Loosely, *luma*. See *luma*, above.

MAC

Multiplexed Analog Component. A color standard that transmits three color components, usually $Y'P_BP_R$, in time-compressed serial analog form. Usually, P_B and P_R are compressed a factor of 2 with respect to Y'. Used mainly with 625/50 scanning.

Metameric

In color science, having different spectral power distributions but identical tristimulus values (*i.e.*, appearing to be the same color): The condition that two spectra, when weighted according to the three spectral response curves of the CIE standard observer, produce the same tristimulus values. Metamerism frequently holds for spectra that are markedly different. Two objects can be metameric with respect to human vision but produce different sets of RGB components when scanned with a television camera that has spectral response that departs from the CIE curves, so metamerism is a practical problem in television camera design. A camera has metamerism errors when it "sees" color differently than a human observer.

Normal line sync

Normal line sync in 525/59.94 and 625/50 systems is a line sync pulse that remains at sync level for about 4.7 µs. In interlaced systems, equalization and broad pulses during the field sync interval are utilized as line syncs, but have a different duration than *normal* line syncs.

Notch filter

A composite color decoder that separates chroma from the composite color signal by a simple bandpass filter centered at the color subcarrier frequency. Notch filters introduce dot crawl artifacts into pictures that have luma detail at frequencies near the color subcarrier.

NTSC

National Television Systems Committee

1. The group that in 1941 established 525-line, 60.00 Hz field rate, 2:1 interlaced monochrome television in the United States. Now properly referred to as *NTSC-I*.

2. The group, now more properly referred to as *NTSC-II*, that in 1953 standardized 525-line, 59.94 Hz field rate, 2:1 interlaced composite color television signals in the United States.

3. A method of composite color encoding based on quadrature modulation of *I* and Q color difference components onto a color subcarrier, then adding the resulting chroma signal to luma. Used only with 525/59.94 scanning, with a subcarrier frequency of $^{455}/_{2}$ times the horizontal line rate, *i.e.*, about 3.579545 MHz.

4. Often used incorrectly to denote 525/59.94 scanning.

NTSC footprint

The first time that the luma and modulated chroma components of an image are added together into a single *composite* signal, cross-luma and cross-chroma artifacts become permanently embedded: subsequent decoding and reencoding cannot remove them. The permanence of these artifacts is referred to as the *NTSC footprint*.

NTSC-4.43

A degenerate version of NTSC, having 525/59.94 scanning and NTSC chroma, but modulated onto a 4.43 MHz color subcarrier instead of 3.58 MHz. NTSC-4.43 is utilized by some European consumer equipment to play NTSC tapes. Provided the lock range of the scanning circuits encompasses 525/59.94, the use of a 4.43 MHz color subcarrier eliminates the second subcarrier oscillator that would otherwise be required.

Odd field
: In a 2:1 interlaced system, a field that begins with a broad pulse coincident with line sync.

Offset sampling
: A digital video system in which the luma samples of a frame are arranged spatially such that the samples of one line are offset one-half the sample pitch from the previous line. Also known as *quincunx sampling*. Contrasted with *Orthogonal sampling*, below.

Orthogonal sampling
: A digital video system in which the luma samples of a frame are arranged spatially in a rectangular array. Contrasted with *Offset sampling*, above.

PAL, Phase Alternate Line
: **1.** A composite color standard similar to NTSC, except that the V-axis subcarrier reference signal inverts in phase at the horizontal line rate, and *burst meander* is applied. Commonly used in 625/50 systems with a subcarrier frequency of 4.433618750 MHz, but also used with subcarriers of about 3.58 MHz in the PAL-N and the PAL-M systems.

: **2.** Often used incorrectly to denote 625/50 scanning.

PAL footprint
: See *NTSC footprint*, on page 289.

PAL-M, PAL-525
: Formally, M/PAL. A composite color video standard employed in Brazil, having 4.2 MHz video bandwidth, 525/59.94 scanning, and PAL color encoding using a subcarrier frequency of about 3.575612 MHz.

PAL-N
: Formally, N/PAL. A composite color video standard employed in Argentina, having 4.2 MHz video bandwidth, 625/50 scanning, and PAL color encoding using a subcarrier frequency of about 3.582056250 MHz.

P_B, P_R
: Scaled color difference components, blue and red. Versions of blue minus luma $B'-Y'$ and red minus luma

$R'-Y'$ scaled for unity excursion (±0.5) for component analog transmission. P_B and P_R are comparable to C_B and C_R, scaled by the factor $^{219}/_{224}$. See also *CB, CR*, on page 276, and *U, V*, on page 297.

Peak white

A term used loosely to mean *reference white*. Reference white is not usually "peak," because analog and digital video systems typically allow luma to excurse to a peak considerably above the reference white level.

Pedestal

Black level expressed as an offset in voltage or IRE units relative to blanking level. Conventionally about 54 mV (7.5 IRE) in ITU-R System M, SMPTE 170, and the archaic EIA RS-343-A; conventionally zero in all other systems, where blanking level and black level are identical. Pedestal is properly an offset; it is incorrect to express pedestal as a percentage. See also *Setup*, on page 295.

Picture excursion

The excursion from blanking to reference white. 100 IRE by definition; $^5/_7$ V (about 714 mV) for System M and the archaic EIA RS-343-A, and exactly 700 mV in other systems.

Picture line

A raster line that may contains picture information, as distinguished from a line that contains vertical sync elements or vertical interval signals.

Picture-to-sync ratio

The ratio between picture excursion (from blanking to reference white) and sync excursion (from synctip to blanking). Conventionally 10:4 for System M and the archaic EIA RS-343-A, and 7:3 in other systems.

Pixel

Picture element. The collection of digital information elements that are specific to a single spatial sampling site in an image – for example, three color component samples and a transparency (*alpha* or *key*) component sample. The term is ambiguous when subsampling is involved.

Power function

A function of the form $y = x^a$. Sometimes confused with *exponential function*, which has the form $y = a^x$.

Gamma correction in television involves a power function often written $y = x^{\gamma}$, where γ, pronounced *gamma*, symbolizes a numerical parameter having a value of about $^1/_{0.45}$.

Quincunx sampling — See *Offset sampling*, on page 290.

R'–Y' — See *B'–Y'*, *R'–Y'* on page 275.

Raster — From the Greek *rustum* (rake): A pattern of parallel horizontal scanning lines that paints out a picture. The raster is the static spatial pattern that is refreshed with successive frames of video. Usually the term is taken to include the sync elements associated with the signal.

Recommendation BT.601, Rec. 601 — More formally, Recommendation ITU-R BT.601. The international standard for studio digital video sampling. Rec. 601 specifies a sampling frequency of 13.5 MHz, $Y'C_BC_R$ coding, and this luma equation:

$$Y'_{601} = 0.299\,R' + 0.587\,G' + 0.114\,B'$$

Recommendation BT.709, Rec. 709 — More formally, Recommendation ITU-R BT.*709*. The international standard for high-definition television studio signals. Chromaticity and transfer function parameters of Rec. 709 have been introduced into modern studio standards for 525/59.94 and 625/50.

Reference black — The level corresponding to picture black. In systems having 7.5 percent setup, such as 525/59.94, reference black is nominally 7.5 IRE units. In systems with zero setup, such as 625/50, reference black is nominally zero.

Reference white — The level corresponding to white, 100 IRE units by definition. Conventionally, white corresponds to light having the color properties of CIE Illuminant D_{65}.

Resolution, limiting — **1.** The spatial frequency, expressed in cycles per picture width or cycles per picture height, at which the response of a video system has fallen to one-tenth of its reference value. Usually, different resolution limits apply to origination, transmission, and display equipment.

2. Horizontal spatial resolution (see 1 above), expressed in units of megahertz.

3. Horizontal spatial resolution (see 1 above), expressed in units of *TV lines per picture height* (TVL/PH, or, colloquially, *TV lines*), where a TV line corresponds to the vertical extent of a single scan line and there are two cycles in a TV line.

RF modulation

In video, a composite video signal that has been modulated onto a *radio frequency* (VHF or UHF) carrier in the range 50 to 900 MHz. RF modulated video in electrical form is usually conveyed with coaxial cable using *type-F* connectors, as in cable TV. A video connection from a VCR to a receiver is conventionally RF modulated onto channel 3 or channel 4.

R, G, B

Strictly, red, green, and blue tristimulus components (linear-light). Loosely, red, green, and blue nonlinear primary components, as explained below and properly denoted *R', G', B'*.

R', G', B'

Red, green, and blue nonlinear primary components. In video usage, gamma correction is implicit: *RGB* tristimulus signals are assumed to have been subjected to *gamma correction*. The precise color interpretation of *RGB* values depends on the *chromaticity coordinates* chosen for the RGB primaries and the chromaticity coordinates of reference white. The FCC 1953 NTSC standard (obsolete), SMPTE RP 145, EBU Tech. 3213, and Rec. 709 all specify different primary chromaticities.

RS-170

EIA Recommended Standard 170 defined the 525/60.00/2:1 scanning standard for monochrome television at exactly 60 Hz field rate. The standard is no longer in use. See also *RS-170-A*, below.

RS-170-A

This term refers to the proposed Revision A to EIA Recommended Standard 170, which was never in fact adopted. The term is used loosely to refer to the timing diagram associated with 525/59.94/2:1 NTSC, as documented in EIA *Industrial Electronics Tentative Standard No. 1* published on November 8, 1977. Had revision A to RS-170 been adopted, it would now properly be referred to as *EIA 170-A*. The recently adopted SMPTE 170M-1994 supersedes all of these.

RS-343-A

This designation refers to the archaic EIA Recommended Standard 343, Revision A, adopted in 1969. It applied to industrial monochrome television systems with 2:1 interlace, 60.00 Hz field rate, a horizontal blanking time of 7 μs, 7.5-percent setup, and between 675 and 1023 lines. Although most of the parameters of RS-343-A are now abandoned, its level parameters – particularly 7.5-percent setup – have unfortunately been inherited by computer equipment.

Sample

1. The value of a bandlimited, continuous signal at an instant of time and/or space. Usually, but not necessarily, quantized.

2. Component.

Scanning standard

The parameters associated with raster scanning of a pickup device or a display device, or the associated signal in a channel or a recording device. A scanning standard is denoted by its total line count, field rate, and interlace ratio, conventionally written as these three quantities separated by a virgule (slash); for example, 525/59.94/2:1, 625/50/2:1, or 1125/60/1:1.

SCH

The phase relationship of *subcarrier* to *horizontal* sync: the phase displacement measured in degrees of subcarrier between a reference point and the closest subcarrier zero-crossing. In NTSC the reference point is 19 subcarrier cycles from the 0_H datum.

SECAM

Séquential Couleur avec Mémoire. A composite color standard based on line-alternate $B'-Y'$ and $R'-Y'$ color difference signals frequency modulated onto a subcarrier. In use only for transmission in certain countries with 625/50 scanning (*e.g.*, France, Russia). No SECAM production equipment is in use; 625/50 PAL or component equipment is used for SECAM production.

Serration

The interval between the end of a *broad pulse* and the start of the following *sync pulse*. This term refers to the absence of a pulse rather than the presence of one, and is deprecated in favor of the terms *equalization*, *broad*, and *normal* sync pulses.

Setup

Black level expressed as a percentage of the blanking-to-reference-white excursion. Conventionally 7.5 percent in System M, and the archaic EIA RS-343-A. Conventionally zero in all other systems, where blanking level and black level are identical. Setup is properly expressed as a percentage; it is incorrect to express setup in voltage or IRE units. See also *Pedestal*, on page 291.

Standards converter

Equipment that converts a video input signal having one scanning standard into an output signal having a different scanning standard with similar resolution. An example is a 525/59.94-to-625/50 standards converter, colloquially known as an NTSC-to-PAL standards converter. See *Scanning standard*, on page 294. *Upconversion* is to a standard with substantially higher resolution; *downconversion* is to a standard with substantially lower resolution.

S-video

An interface that conveys *luma* and quadrature modulated *chroma* separately as two signals on a specific four-pin mini-DIN connector. There are two only types of S-video: *S-video-525*, which has a 525/59.94 raster and chroma modulated with a 3.58 MHz subcarrier, as in NTSC; and *S-video-625*, which has a 625/50 raster and chroma modulated with a 4.43 MHz subcarrier, as in PAL. S-video uses quadrature modulation but not frequency interleaving. It is not exactly component video, and not exactly composite. Most S-VHS and Hi-8 VCRs implement the S-video interface.

Sync

1. The process of synchronizing the scanning of receiving, processing, or display equipment with a video source.

2. A signal comprising just the horizontal and vertical elements necessary to accomplish synchronization.

3. The component of a video signal that conveys horizontal and vertical synchronization information.

4. *Sync pulse.*

5. *Sync level.*

Sync level

The level of a sync signal or component. Sync level in a composite video signal is conventionally -40 IRE (-285 $^5/_7$ mV) in System M and the archaic EIA RS-343-A, and -300 mV in other systems.

Sync pulse

A normal line sync pulse, equalization pulse, or broad pulse.

Synctip

The level or duration of the most negative excursion of a sync pulse from blanking level.

System M

Formerly *CCIR System M*; now properly referred to as *ITU-R System M*. A television system having 525/59.94/2:1 scanning, 4.2 MHz video bandwidth, and 6 MHz channel spacing. The designation does not indicate color encoding. Most color 525/59.94 transmission is System M/NTSC, although Brazil uses System M/PAL.

Timecode

A number of the form *HH:MM:SS:FF* (hours, minutes, seconds, frames) that designates a single frame in a video or film sequence.

Transcoder

A circuit or equipment that converts a video input signal having one color encoding method into an output signal having a different color encoding method, without altering the scanning standard. An example is a PAL-to-SECAM transcoder.

Trichromaticity — The property of human vision whereby mixtures of exactly three properly chosen primaries are necessary and sufficient to match nearly all colors. This is surprising considering that physical spectra are infinitely variable, but not surprising considering that the retina contains just three types of photoreceptor *cone* cells.

Tristimulus — A signal that represents radiant power (or, loosely, intensity), weighted by a spectral sensitivity function having significance with respect to the *trichromaticity* of human vision (see *Trichromaticity*, above). A tristimulus signal represents linear-light, and is subjected to *gamma correction* as part of its conversion into a video signal.

TV line, TVL — In a television system, a unit of resolution having a dimension defined by the pitch between two adjacent scan lines. In an interlaced system, adjacent scan lines are in opposite fields. Horizontal and vertical resolution are commonly measured in units of TV lines, where there are two lines per cycle. See *Resolution*.

U, V — **1.** Color difference components, blue minus luma ($B'-Y'$) and red minus luma ($R'-Y'$), scaled by the factors 0.492 and 0.877, respectively, prior to quadrature modulation, in order to contain the excursion of the composite color signal within the range $-33\frac{1}{3}$ IRE to $+133\frac{1}{3}$ IRE. See also C_B, C_R; I, Q; P_B, P_R.

2. The symbols U and V are sometimes used loosely or incorrectly to refer to unscaled $B'-Y'$ and $R'-Y'$ components; to components P_B and P_R that are scaled for component analog transmission; to components C_B and C_R that are scaled for component digital transmission; or to color difference components having unspecified, nonstandard,or unknown scaling.

Unit — See *IRE, IRE unit*, on page 286.

VCR — Videocassette recorder. Implicitly, consumer-grade. See *VTR*, on page 298.

Vertical blanking	That number of scan lines in a field that are mandated by a raster standard to contain nothing except field sync elements and blanking. Note that only a small number of *vertical interval* lines are actually blanked – nine in the case of NTSC. The original FCC, NTSC, and EIA standards interpret vertical blanking as the time interval between the end of picture information on the bottom picture line of one field and the start of picture in the following field; modern usage refers to this as the vertical *interval*, counts integral scan lines rather than microseconds, and recognizes that the vertical interval may contain nonpicture video.
Vertical blanking interval, VBI	See *Vertical blanking* and *Vertical interval*. This is an ambiguous term because not all of the vertical interval is blanked. Usually VBI refers to the entire vertical interval, including signals such as VITS and VIRS, and excluding only the picture lines. Sometimes VBI refers to just vertical blanking.
Vertical drive, VD	A pulse of the same duration as vertical blanking that conveys vertical synchronization information. Nowadays this information is usually extracted from *reference video,* rather than requiring a separate signal.
Vertical interval	Those raster lines that are precluded by a raster standard from containing any picture content. Vertical interval lines that are not mandated to be blanked can be used to convey test signals (VITS), a reference signal (VIRS), timecode (VITC), closed captioning data (CC), teletext, or other information.
VTR	Videotape recorder. In broadcast (studio) usage, *T* for *tape* is used even if the tape medium is encased in a cassette.
White	See *Reference white*, on page 292.
Y	In physics and color science, the symbol for linear-light *luminance*, the *CIE Y* tristimulus value. In video and computer graphics, often confused with *luma*.

Y'

In television, the symbol for *Luma*: the sum of nonlinear (gamma-corrected) red, green, and blue primaries, each weighted by its *luma coefficient*. Historically the *Y* symbol in video was *primed* (*Y'*), but the prime has been elided in modern times, leading to widespread confusion with *luminance*.

Y'/C, Y'/C3.58, Y'/C4.43

Luma accompanied by a single *chroma* signal, quadrature-modulated at approximately the subcarrier frequency indicated in megahertz. More properly referred to as *S-video, S-video-525, S-video-625,* respectively.

Y'/C629, Y'/C688

Luma accompanied by a single *chroma* signal, quadrature-modulated at approximately the subcarrier frequency indicated in kilohertz. Used as an interface standard for dubbing, editing, or timebase correction of certain *color-under* VCRs.

$Y'C_1C_2$

1. A component color system conveying luma and two color difference signals, where the components C_1 and C_2 are evident from the context and are not necessarily any of the common pairs $B'-Y'$, $R'-Y'$; C_B, C_R; P_B, P_R; *I*, *Q*; or *U*, *V*.

2. $Y'C_1C_2$ (1) according to Kodak's PhotoYCC coding.

$Y'C_BC_R$

Luma accompanied by scaled color difference components. In analytic use, the components are scaled independently so that each has an excursion of -0.5 to +0.5, relative to a luma excursion of unity. In a digital electrical interface, C_B and C_R are defined by Rec. 601 to encompass codes 16 through 240, with respect to 8 bits. This corresponds to an excursion of the fraction $^{224}/_{219}$ of the luma excursion. See *Luma, Luma coefficients*, on page 287; and *CB, CR*, on page 276.

Y'IQ

Luma accompanied by color difference components *I* (In-phase) and *Q* (Quadrature). *I* and *Q* are derived from *U* and *V* by a +33° rotation and an exchange of axes. Historically NTSC was formed from color difference components of unequal bandwidth, where *Q* was filtered more severely than I. Nowadays quadrature modulation based on equiband color differences is almost universal and $Y'IQ$ is obsolete, except for $4f_{SC}$ NTSC composite digital interface and D-2 composite digital VTRs.

$Y'P_BP_R$

Luma accompanied by P_BP_R color difference components in analog form, scaled to have the same peak-to-peak excursion as Y', typically -350 to +350 mV. See *Luma, Luma coefficients*, on page 287; and *PB, PR*, on page 290.

$Y'UV$

Luma Y' and two color difference components U and V scaled appropriately for encoding into *composite* video. See *Luma* and *U, V*. $Y'UV$ is useful only toward subsequent encoding of the components into a composite video signal, such as NTSC or PAL, and is not appropriate for representation of component video. The notation YUV is often used loosely to denote any component system employing luma and two scaled $B'-Y'$, $R'-Y'$ color difference components.

Index

C

D

E

F

K

L

O

P

S

T

U

V

W

X

Y

Z

This book is set in the *Syntax* typeface, designed by Hans Eduard Meier and issued by Stempel in 1969. The body type is 10.5 point, set ragged right and leaded to 13.2 points.

The mathematical work underlying this book was accomplished using Wolfram Research's *Mathematica*. The illustrations were executed in Adobe's *Illustrator*; for raster (bitmap) images, Adobe's *Photoshop* was used. The equations were set using Design Science's *MathType*. Text editing and typesetting were accomplished using Frame Technology's *FrameMaker*. Adobe's *Acrobat Pro* was employed for electronic distribution.

The work was accomplished using Apple Macintosh and Sun SPARCstation computers. Proof printing was done on Apple LaserWriter Pro 810 and Xerox DocuTech 135 printers. The electronic master was produced using Adobe's *PostScript* language.